Anna's Ossam Adventure

a Himalayan road trip
in the Shadow of the Devi

To Julia, a Woolmer! and part of the story! with love Nigelle, Zoé + Sita Devi! xxxx

Copyright © Nigelle de Visme 2019

Nigelle de Visme has asserted her right under the Copyright, Designs and Patents Act 1988 to be identified as the author of this work.

This book is sold subject to the condition that it shall not, by way of trade or otherwise, be lent, resold, hired out, or otherwise circulated without the publisher's prior consent in any form of binding or cover, including this condition, being imposed on the subsequent purchaser.

Photograph credits: thanks to AK for selfies and photographs of Anna (me) and thanks to NdeV (me) for all photographs of AK and views; Chelça Thurlow for page 37. The remarkable photo of the light over the icons on page 49 is by Ann Cook FRSP and used with grateful permission

The poem Walk in Beauty © Samantha Reynolds is used with her permission and included here with love

Anna's Ossam Adventure

a Himalayan road trip
in the Shadow of the Devi

Nigelle de Visme

Other titles by Nigelle de Visme

The Tiger and the Taxi Driver – Book One
(Memoir)
The Tiger and the Taxi Driver – Book Two
(Memoir)
Patrick and the Cat Who Saw Beyond Time
(Foreword by Bede Griffiths and with coloured illustrations)
The Road to Delphi – a gentle murder
Same Skirt Different Day – A Camino Chronicle
Anna's Ossam Adventure – a Himalayan Road Trip

Dedication

To A.K. – my world for a whole three weeks – ossam!

Arun is not his real name, nor is Anna her real name. Anna is me, but the true identity of my wonderful taxi driver will remain a secret. I will call him Arun after the charioteer of the sun-god Surya, for he was indeed my charioteer. He fought battles for me, thought for me and even bought for me. He was the most caring of men, and I could not have found better. Heartfelt thanks! As for wise great aunt Anastasia, she is really Mrs Tweedie, my beloved Teacher in whose company I sat, as Indians say, for four years absorbing all I could to put in my spiritual dillybag and continue my Journey.

Foreword

I have chosen to publish this slightly rearranged version as a stand-alone story from my memoir The Tiger and the Taxi Driver as a gift to the real taxi driver who looked after me so well during our Himalayan road trip in October 2018. We explored, for three weeks, the Kumaon foothills in the Shadow of the Devi. It was a pilgrimage for me, one which I had longed for on the periphery of my mind for forty years. It never entered my dreams as the remotest possibility before the advent of a mobile phone in my life at the late age of seventy-two. Whatsapp sealed my fate, the fate of the taxi driver and a pilgrimage of colour-filled dreams wove me a Lavender Codicil.

Walk in Beauty

I am not old... she said
I am rare.
I am the standing ovation
at the end of the play.
I am the retrospective
of my life as art
I am the hours
Connected like dots
into good sense
I am the fullness
of existing.
You think I am waiting to die...
but I am waiting to be found
I am a treasure.
I am a map.
And these wrinkles are
imprints of my journey.
Ask me
Anything.

Samantha Reynolds

Contents

1. I Come in the Little Things
2. From the Tiger to the Taxi Driver
3. A Lavender Codicil
4. The Devi Calls – on WhatsApp
5. Ossam, just Ossam ...
6. "Welcome, Distinguished Guests"
7. Coming Home
8. Bathos!
9. Tibet! Tibet!
10. Peter Rabbit
11. Miracle of the Moon
12. "Ma Ganga Is Always Pure"
13. Delhi Daze
14. Carp Whisperer

Ustad – the Great Being who began it all

1

I Come in the Little Things

Words of power and prophecy had embedded themselves in the very cells of Anna's little ten-year-old body: *I don't want to live on this earth when the last tiger has gone,* she told her mother. *The child is mad,* was her mother's opinion of that; and, *All the best people are a little mad,* said aunt Alice of herself and of Anna.

Sixty years later Ustad's spectacular tiger beauty filled facebook pages for months and woke the memory of Anna's prophecy: *Like a sleeping tiger,* she said to friends.

Anna asked to join a small group of Indians who had booked a photographic safari with a tour leader from Uttarakhand. Anna had found out about it from facebook; they were going to Ranthambore where Ustad once roamed. She had to go, had to make a sacred footstep and leave her prayers in Ranthambore in small atonement at the iniquity of the human race for their destruction of entire species, vast forests – well, pretty well the whole world, during Anna's lifetime.

Ustad, like Blaze, the Bear of Yellowstone, where Anna had gone in atonement at *her* dreadfully contentious death from human stupidity, to lay prayers in the forest of the Grizzly, was a symbol. Cecil the lion was another singular sacrificial symbol, and, though Anna felt no call to Africa, she prayed silently in the solitude of her home for the great lion wounded, by how many arrows? and left to die over days in unspeakable agony, shot at by a trophy hunter dentist from Minnesota. Did these individual Beings, named and known, *choose* their suffering to alert the human race to the anguish of their passing? A passing that pointed to their whole species? It was a thought Anna pondered deeply.

Anna contemplated and on cosmic cue unrelated tiger synchronicities appeared to convince her she should go to India, all would be well and her prayers for the Tribe of Tiger would be registered, Somewhere. A circlet of small miracles quickened her wavering intent. It was more than Tiger she would meet.

Ustad. The majestic tiger of Ranthambore. He had, so it was alleged but unconfirmed, killed one of the forest rangers and the decision about his fate split the conservation community of the world. Each wanted the best for Ustad under such harrowing circumstances but what the best could be when no other tiger reserve in India would take him would become a dilemma of tragedy, no matter which final decision won. Images of Ustad filled the facebook pages of many links Anna had added to her own page and each showed Ustad as so handsome, so majestic, so dignified that Anna knew he was a special Being; a Being whose attention was bringing focus to the reality that the tribe of tiger was tottering on extinction.

Simultaneously some of the better newspapers of the day were revealing the ever increasing depravities of Thai Buddhists farming tiger cubs and drugging adult tigers for tourists to fondle, to faddle, reducing Tiger to a trifle, a cipher, a selfie. Thirty-three dead tiger cubs bred and farmed for body parts to sell to the Chinese were found stuffed in the Temple's freezer when the police finally had enough evidence to pounce and prosecute.

Through the mists of Mnemosyne words of Anna's own prophecy sounded, faint and far as a star, for stars emit sound, and the Bushman mother sings: *Can't you hear the stars? The hunters? Tsik! Tsa! call their voices. And the greatest Hunter of All is never seen for He is in the Darkest Space, hidden. O, take the heart of my child and give him the Heart of a Hunter...*

Over a long life of mostly solitude Anna had heard mountains sing and stars sound. When she learned a woman had discovered pulsars the knowledge thrilled her viscerally. Jocelyn Bell-Burnell told the world all was made from the same elements as stars: carbon, hydrogen, iron; surely the same vein pulsing along the boundless *Yes!* that governed Anna's life. No Nobel prize for Jocelyn and her pulsars though, it was given to her male supervisor. Her name? That took a long time for recognition. Anna's long journey to herself would come to embrace a womanroar of righteous anger at the appropriation by multitudes of men of all the wonders of women's minds and of their debasement of women's bodies. *Women*, whispered her awakened Womanroar somewhere in the late 'eighties, *are wonderful,* and feisty writers the like of Susan Griffin and Sonia Johnson, with a nod to Mary Daly, informed her, shored up her growing unease that equality was only equal for some. After her hysterectomy she woke with a roar where once was her womb – a *Womanroar*.

Anna knew the stars who made her. She had heard stars breathe in their never-ending slow, deep rhythm, watching as worlds come and go. She was made of stars, of stardust, of iron and silver, gold and carbon, a shower of elements that rained through the galaxies when stars imploded billions of years ago. Stardust was in her mitochondria. Long ago in Malaya Asmah, her amah, had taken her out at night to search the heavens and say, when her parent's quarrels rent the air: *Don't cry Baba, look up, we come from the stars and to them we return.* It was a big think, but Anna was big even when she was a small child. How else was prophecy possible from a ten year old who remembered the stars who made her? How else would a ten year old know that in her own lifetime the majestic Tiger would become extinct?

Anna *did* want to see a tiger before they, and she, left the planet. Her heartbeat and the talisman of Signs told her she *was* going to India and all would be well.

Ranthambore was a long way from Corbett National Park yet despite numerous hiccups and horrors all would fuse into an amalgam of remarkable experiences that Anna's return to India: *After an absence of more than quarter of a century,* she marvelled after the event, had opened her to. The greatest blessing was Tiger, she *had* seen tigers, in Ranthambore and in Corbett, meetings pared to essences, close to, majestic, holy. Her pocket camera was enough. Her prayers for Ustad remained in the air, there.

Anna included a few days in a private wildlife reserve outside Corbett National Park owned by a friend of the facebook safari contact, a young Bollywood songwriter mogul who had bought a corridor of forest for the big cats in the Kumaon Hills. *Kumaon Hills!* The words thrilled Anna. Those hills were the setting for the forest of the Heavenly Horse written more than thirty years before. She had never been there. Anna's skin tingled, an eerie, cellular, response to *Something*.

It is the mythical forest where Sita was abducted by Ravanna; you will be completely alone, she was told.

A *frisson* at the thought of being alone in a forest in the foothills of the Himalayas was momentarily daunting. Anna would not be able to speak a word of any Indian language. But – what else but *Yes!* could Anna say to solitary in Sita's own forest?

The braided challenge of solitude and singular worldly failures made her smile as she emailed *Yes* back to Lata, the facebook friend; Anna's life had inclined her to solitary. Later, Lata's promise of being 'alone' would prove an Indian fiction. As for the vagaries of food, the timing of which was her life support in place of metformin, *that* variable would put her perilously close to an insulin coma once Lata disappeared over the horizon leaving Anna in a wild place many walking kilometres from the nearest, barest, store and a mile down a goat track to the nearest passing car. Extravagantly written instructions of food and its timing for her meals and given to the two hill boys who couldn't, or wouldn't, read, discharged Lata's duty to Anna.

The forest retreat was a visual charm. A small cottage built on a small rise within the forest, its downside was too many rooms for privacy or silence. Anna had two nights of pure solitude before a delightful but disrespectfully loud family arrived from the din of Delhi, bringing it with them. Occupying the other rooms in the tiny building their loud voices, their radio with its inappropriately loud film music, powerful torch light seeking to penetrate the jungle night, gave lie to Lata's promise of solitude; a solitude which did not come cheap – Anna was a foreigner, rates for her were set accordingly. She didn't mind paying to support the leopard, the munjac, the blue tailed magpies, the water springs – but she minded acutely paying for Delhi din.

Her initial twenty-four hours of solitude brought magic. Madan, one of the two young men who lived onsite, took her for a walk along the adjacent hill path, waiting as she negotiated the rocky steps, pausing as she caught her breath. At a point he indicated they should leave the path to take a small divergence which brought them to an ancient tree and an even more ancient shrine of slates at its base in a half circle with Sanskrit lettering. Mnemosyne had long since stolen Anna's Sanskrit and she turned to Madan, questioning with gesture the words on the slate.

Sita Ram, he responded.

Sita Ram? Sita Ram! Anna hadn't remembered even this much Sanskrit? The warmth of the day didn't stop her body becoming an amalgam of goosebumps as her soul-name surfaced from its long resting. Sita Devi. This shrine to Sita and Rama, aeons old, in this very forest, kept alive by the daily puja of a priest at dawn down all the centuries since Sita was abducted by Ravanna in the mists of myth rooted in this very forest of Sitabani, was the reason she had been brought *here*. Anna had come full circle. She wished she could, like Penelope Chetwode was said to have done on the steps of one of her beloved temples of Himachal Pradesh to die a death more poetic than any poem written by husband John Betjeman, sit right down and die there.

How, she wondered, could Madan ever know what his unexpected diversion would mean to her? They had no common language and the story too complex to tell by sign. Would he even be interested? She would never know, but the following afternoon he invited her again to walk with him. New quiet guests had arrived to occupy the last of the lower rooms, a medical consultant from Delhi and a High Court lawyer from Calcutta, and Madan was taking the gentlemen from Calcutta for the same walk to the mountain top. This time Anna turned left at yesterday's diversion in order to share the story with her companion whose English was impeccable. Lost in thought, she did not see him take one photo as she strode away from the shrine; a tessera to add to her mosaic of memories.

On her first evening, while she was still alone, Madan had come over to the cottage and gestured for her to follow him onto the terrace. Moonlight pooled the wild garden, creating a chiaroscuro of the high forest beyond. A cough sent a thrill through her, no ordinary sound and no human either. *Leopard*, said Madan simply. The hair rose on Anna's nape, thrilling. That night she woke to a sob, a sobbing cough so close, so soft, it beckoned her, a primal call. Anna left her bed, Leopard must be at her window, but by the time she had risen, gone out through the cottage door and rounded the corner of the small building to the terrace, he had left her, silently slipping back into the forest. Anna was left alone to stand and wonder at such intimacy of a wild creature.

Her mind called up a memory. In the seventies she had visited Bristol Zoo. One moment filled her known world with an inchoate truth: Man! Men had been responsible for: *Jenny, panther pardus, aged fourteen.* Fourteen *years* the leopardess's eyes had burned with longing; now they pierced through Anna to the freedom she would never regain. Her cage was barely fifteen feet by six feet. She paced. She paced. She paced. Around and around until Anna's heart broke. There was nothing Anna could do; in those far off days she didn't know how to pray, how to say: *I am your witness.* From the Pliocene age until now this sublimely beautiful cat had lived, free. And now, if she wasn't behind bars, she would likely be skinned as a coat for a woman. Man and woman, equally guilty. Anna never visited another zoo but Jenny and the green-eyed Fishing Cat she had seen as a child with great aunt Anastasia looked out from Anna's heart forever at the world they had lost. *Now*, she could send a prayer for Leopard.

In the morning she put on the earrings of Hannah Willow and wore them all day. A large silver oblong of a golden hare leaping over the Tree of Life had so touched her she bought them. On their reverse Hannah had engraved: *I will run and run forever where the wild fields are mine.*

Anna's connection with animals was sourced from the same crucible that gave birth to the wisdom of Dom Robert: *La nature est la vraie réalité. La nature ne trompe jamais. La nature est le visage de Dieu – Nature is the true Reality. Nature never deceives. Nature is the Face of God.* She recalled a documentary which broke her heart yet again, as animal truths always did. The documentary of 7th January 1997 had a poignant sub-title: ... *and in the decimated forests the shamans weep:*

Wildlife filmmakers had set up a hidden camera in the decimated forests of eastern Russia near the Altai mountains. It was focused on the cave of a male leopard. Within the cave was a cache of meat. The camera was triggered by movement. Whenever the leopard returned, his behaviour was filmed.

An old female leopard and two cubs came one day when the male leopard was away. They were forced from their home territory by deforestation, man's predation, and the loss of deer.

Leopards are territorial. The leopard would never usually betray the forest's first code of courtesy. But hunger is terrible. The old leopardess went slowly to the cave. She smelt the meat, walked around the cave entrance, gave small coughs of hunger, did not eat; did not call her cubs. All was recorded by the dated eye of the hidden camera. She left. But hunger is terrible.

Ten days later the female and her two cubs came again. Now they were starving. And hunger is a terrible thing. Starvation led them to cross over the forest's codes of courtesy. Starving, the old mother leopard reached the cache of meat in the cave. She ate. Only a little. Then she carried a small carcass from the cave to higher ground. Then, almost inaudibly, she called her cubs. Short, soft and deep-throated, 'wuf, wuf'. Some time passed. These last days the male leopard had not returned to his cave. Suddenly movement triggered the camera's focal range to the left. The male leopard. Approaching so cautiously, as yet unseen by the old mother and her cubs, who were still eating. He padded, unsure, up to and passed the hidden camera. We, the watchers, with our unchaste eyes, our culpability held in common, were given no reprieve from the anguish in the deep pools of amber astonishment that pierced the camera lens and impelled into our unsuspecting hearts.

He paused, as if doubting his own sight. We, the watchers were now spared the bewilderment apparent in his eyes. Now, his face had passed beyond the camera's lens and we could watch him approach his precious cache, his life. He paused again. Then he growled – and the old mother with her cubs sprang away.

The male leopard watched them go. He walked to the cave. He sniffed and looked all around the cave. His meat had gone. There was nothing left. And the leopard sat down and wept. The hidden camera recorded his anguish, but not our aching hearts. And the leopard cried. The cameraman knew the hours recorded on the film...the leopard had cried all night.

There was no more food, no more deer, no more rabbit, no more forest... The village shaman also wept. Our leopard, *he said,* is shedding tears. Our leopard is the icon for our spiritual survival. Our leopard is shedding tears, for our world and his are passing. The amber leopard of Asiatic Russia is an icon, *said the villagers,* for our spiritual survival. We are going ...

Anna wept then. How could she justify her life against such enormity; the iniquity of which she, too, must own as human transgression?

It was time to leave Sitabani. The noisy, pleasant, jolly family from Delhi was too much for Anna at such close quarters, kindly though they were, well-meaning and far more hospitable than She-Who-Walked-Alone through the Wildwood. She arranged to leave earlier than planned with the two quiet gentlemen guests who had arrived so fortuitously, and who, even more fortuitously, were returning to Delhi.

Being there had been a privilege. Forests had their own language, wildlife is Life, the Tiger and the Leopard are guardians of the Rivers. If rivers are clean, animals can drink, if not they will move on, move away, the Forest's principle predators are the highest indicators of its health and well-being. Anna knew her time in the forest had been acknowledged, *Somewhere*. Leaving now was in the right order of things.

When I walked down the garden path as a two and a half year old, carrying my woollen pony through that garden gate, Anna wrote in her journal on her final night in the forest, *I knew that the World 'out there' would love me. It was the people behind me in the house I was waddling away from on my chubby little legs who didn't. We didn't get far, pony and I, just to the vicarage wall at the edge of our cul-de-sac before we were hauled back. No matter, it was an epiphany moment. Out beyond the garden gate the world welcomed me, it held promise and I would love it passionately. Life in my childhood home was rent with rancour ... But the world beckoned. My soul and my feet had been set on a pilgrim way. I was born to a lonely path, yet ...*

... yet She walks with me in fields and forest. I know Her in favourite cups, favourite spoons, wrote Anna, reflective now, *in roses and love-in-the-mist, in reeds and weeds, in paintings I love, clouds, the complete dependability of my old, old car, an old exquisite desk I once had, my two-seater Knole, my chosen doors, wind, trees,*

old stones, cats, woods, sea. I see Her Hand in every beautiful thing. I hear Her Eternal Sigh surrounding me. I am 72 in a minute and She is near, always, in all beauteous things. When She invites me to follow I am Obedient. My sovereign soul, beyond all my myriad smallnesses, has been my Silent Witness through All Time.

Anna's long Quest for Sita had been accomplished. And that, for Anna, was exactly as it should be. The following morning she left with the two gentlemen, sharing the cost of the private car for the five hour journey back to Delhi. They knew of a small hotel in Saket, it would surely have a room *on spec* until she could return to the hotel in Connaught Place expecting her three days hence. The hotel did have a room, reasonable, and there the two men left her, with an invitation to lunch on the third day.

with the woollen pony

2

From Tiger to Taxi Driver

Angels, beautiful pauses in the whirlwind,
Be with us through the seasons of unease.
... Surprised by angels ...
the beautiful pause ...
Opens the way into the sacred dance.

May Sarton – Beautiful Pauses

Anna was waiting in the hotel reception of the hotel in Saket for the promised driver to collect her and drive her to their lunch appointment in his home not five minutes walking distance. Her host had insisted on sending his car at eleven o'clock, the agreed time. Anna had to be firm about eating at midday to avoid the diabetic sugar-crisis that caused collapse.

The hotel phone rang; the car would be fifteen minutes late. At eleven-fifteen the phone rang again, the car would be delayed a further fifteen minutes. Anna, unsustained by the frugal breakfast offered by the hotel, had walked to the Coffee Shop at ten-thirty for a beetroot and ginger 'Goddess Elixir'. Though hardly substantial enough to keep her sugars balanced for very long she had no reason to doubt the luncheon arrangement for midday. At ten minutes to twelve, her witching hour, she gave up, she was losing her composure, her shaking hands warning her that she had to eat, fast and within minutes; insulin coma a serious possibility. The car her host was sending to collect her could continue its fifteen minute increments of delay indefinitely and who knew what kitchen crisis might meet her once she reached her destination. She asked the receptionist to call a taxi, *now*, she would leave, *now*, to Connaught Place; Zaffran was on the same floor as her hotel in D Block, she would eat lunch there, food would be ready, it was so close to midday. She prayed for a Godsend.

Within a minute she was in a taxi, safe, her need to eat now her own responsibility. The taxi turned, fed back into the northbound traffic – and gridlocked. The sea of traffic ahead of them was motionless, five lanes abreast, the taxi was trapped. The driver's phone rang – Anna knew it was her host, the taxi driver's number passed on to him by the hotel receptionist. What could she do? In her shaky state the only thing she was sure of was getting to Zaffran; even could they turn back, she wouldn't. The young driver answered in Hindi, fending off the caller, Anna was sitting next to him in the passenger seat. Tapping the driver lightly on his arm he looked at her mouthing: *No!* and gesturing with her hand that they were to continue. The driver spoke into the phone in abrupt monosyllables, hung up. Moments later the phone rang again, the scene repeated. Anna didn't want to speak to anyone, the stress now they were gridlocked and stationary seriously beginning to tell.

Who is this man? asked the driver of Anna, between fending off harassing phone calls. Her explanation of: *Medical consultant* cut no ice. *He doesn't sound like one, I think it is his driver phoning me and telling me to take you back.* Something in the young man's proprietorial manner made Anna smile but when his phone rang a fourth time the young driver ignored Anna's refusal to speak and handed her the phone. This time it was her host, adamant her taxi turn back. *We can't,* said Anna, focused only on food somewhere at the end of the hellish journey, *we are gridlocked, in the centre lane of five unable to move.*

Her host, properly upset but refusing to acknowledge his tardiness was likely responsible for having ignored Anna's needs – *and he a medical man,* she could laugh later – suddenly said: *Your taxi driver, he is a most unscrupulous man, I want him to bring you here, now!*

At such an inappropriate adjective Anna blazed into defence: *My driver is absolutely scrupulous,* she said, a shard of ice cooling the heat of the day and their discussion. *And you know I need to eat, so I must continue on to the restaurant. I am so sorry this has happened, but food on time is my medical support and you of all people should know this.*

Food is ready, he said briskly, and her response: *but I couldn't know that, your delays could have continued all afternoon, how could I know?* ended the call with regret on Anna's part that such a thing had happened at all.

She handed back the phone and sat trembling now the witching hour of midday was upon her. Her young driver glanced at her. The taxi inched forward, the driver spoke: *My mother,* his voice was quiet, *has diabetes, she goes like you when she doesn't eat on time. Are you diabetic?*

Anna nodded, adding how concerned she was that it would be, according to her gallant, perhaps an hour before they reached Connaught Place in such traffic. *You must be calm, relax,* came his answer, *I will get you there, do not worry, trust; the body does not understand, even as the mind does; be calm, God will keep you safe.*

His English was of the endearing kind and his words washed a balm of calm over her: *Like a charm*, she was to say later, *as if God had lifted all the stress and dissolved all those sugar-shakes, just-like-that. That taxi driver* was *a Godsend.* She reflected that some people she had known had healing in their very presence, a gift in those who were unaware of it. Like this young taxi driver.

Anna sat back, shoulders dropping from their held tension, looking about, stealing a glance at the man next to her whom she had barely registered till now. *Lovely mouth*, she thought, *gentle*, she could trust this man. As they inched along, spurts of progress hampered by all the traffic of Delhi streaming into its centre, he tore off a sheet of paper from a notepad tucked behind his windscreen shield, wrote his name, his phone numbers, and passed it to her. She sounded his name: *Arun*. Smiling to herself she clocked the name as that of Surya's charioteer; she was in good hands.

The long journey in to the city centre occasioned little conversation other than plans for her remaining three days. Anna asked if Arun would come for her tomorrow, she had an invitation to meet someone at the Gymkhana Club, and, the following day, sightseeing in the morning and then to the airport at midnight. He would be most happy to, his assurance a comfort.

They reached the hotel. The fare, added to by the unconscionable traffic delay, was clearly visible on the meter. Anna handed Arun notes and insisted he keep the change. *No,* he said in his firm way, *less, less* and he peeled off more than the fare he was owed to return to her refusing hand: *No, no,* said Anna, *just keep it!* How could she tell him that his patience and his insight over her condition and the harassing phone calls he had fended off were priceless? How would she know that rarely in her life had she been cared for, *seen,* in her frailty and accepted as she was?

Madam, he said softly but firmly, *if you insist I take this I will not come to take you to the airport.*

Stung, Anna's womanroar leapt from its sleeping: *You will!* she said, imperiously, with a few centuries of Empire weighting the delivery of her tone, and then laughed while he eyed her, still refusing to keep what was his due, but nodding a promise with his ghost of a smile, saying: *I will not abandon you.* The word *abandon* buffered Anna's heart, like wild wind on lonely littorals, burst her heart open. She was born to *abandon*; its hollow ring chased all the years since her mother turned from the birth of her own daughter, Anna, more than seventy years before, and her father abandoned the marriage which brought him such unhappiness. He could not save his daughter's life; she had to learn to swim. Anna hurriedly closed her heart, but the suddenness left meltwater in its wake.

And so her last day in Delhi dawned. Anna asked to be taken to gardens, suggesting the Jayanti Gardens to begin with, she wanted to photograph peacocks in their urban splendour – their wilder cousins of Ranthambore were nervously flighty and had not paused for photo shoots. He would come: *Early, early,* he said, *at six-thirty the birds will be there and your peacocks,* but Anna sighed, *Arun, diabetes, I must eat, breakfast doesn't begin until seven, come at eight and I will eat earlier, but if I breakfast too early I will collapse and the morning will be lost.* He came, of course, and her heart lifted as she saw him from the hotel entrance flicking the Delhi dust from *her* seat, wiping *her* window clean of any speck; his white shirt so dazzling the shadowed folds of it glowed a blue hue, his shoes polished to mirror gleam. She noticed his attention to appearances this day, the pleasure in their greeting, mutual.

No traffic slowed them now as he drove to the most romantic garden in the world, the mist of the morning chasing the light haze over its waterways, the bamboo thicket, the vibrant multicoloured kaleidoscope of towering bougainvilleas, the open parklands where the heavy limbs of majestic trees arched serpentine to the ground. She took a photo of him by a wall of deep magenta bougainvillea, his gentle gaze directed to Anna through the lens, his mouth touched with upturned warmth. He insisted on reciprocating the photograph but in hard sunlight unsoftened by shadow or mist Anna inwardly curled in discomfort; the martinet lines by her mouth, a French inheritance, the creviced crease down her forehead puckered deep with decades, the calligraphy of lines marking all the sorrows and losses of her life, a cartography of her known world charted clear on her face, all, *all*, would be etched in high relief in the bright morning. And it was. She could not, of course, have refused.

She didn't know then and there that he felt a bond with *her*, her frailty and her face was part of *her*. Even less did she know she loved him with a troubadour kind of love and probably as ancient. Indians, and the good folk of Glastonbury, would name it an old karma. That garden, the mist, the unfolding beauty as they walked deeper into its wildness; his *'come, come, come'* as she followed his lowered hand curling in the direction of each softly spoken imperative, rewarding each destination with a new vista: a rose bed, an old Sikh collecting fallen petals to fill bowls on his table, a golden Buddha, a monolith engraved with words from HH the Darling Lama, a waterfall, ducks, squirrels with their gratitude stripes from Rama's hand for their help in carrying a million pebbles to build a bridge for Rama to cross when rescuing Sita ... No, she certainly didn't know, then. Did *he* know, or was he normally so kindly? Had she even occasioned the thought would she, would they, have fallen into that mutual silence brought about by an intensity of liking confounded by lack of common language with which to communicate without misunderstanding? Confused by that force from the heart, would they identify the attraction, unbidden, that leaps across every barrier and boundary? Was the fact he was just half her age relevant ... to *anything*?

Anna, after three weeks back in India, was acutely aware of the silent pyramid at the apex of which lived the beautiful, the privileged; and aware that at its bottomless base lived the massed misery of people born to be perpetually hungry, ignored, downtrodden, exploited. Arun was not one of them, he drove a taxi, perhaps he owned it, was proud of his position. She wanted to invite him to lunch with her at the hotel.

When she had handed the slip of paper on which he had written his name and phone numbers to ask the proud little peacock on reception that day to call Arun for her she was told that one couldn't call individual Uber drivers. *Of course you can,* said Anna briskly, *he's been my driver and I like his driving.* Peacock sniffed down his nose: *then he is making money on the side* … Anna forbore from saying the obvious: *And working days of sixteen hours he deserves to have someone like me paying him properly from time to time!* Pointless, too, telling him that Arun had refused to take the correct fare she had offered; *less less,* he had said, peeling off notes to insist she take back.

Peacock had attempted to put her firmly in her own white, post-Independence, place during her first days there. She had brought the brass nozzle of an oil lamp with her to India. The hand-blown glass burner was an amalgam of kingfisher blues and swirling greens made by an artist well-known in Adelaide, its wick had long since burned away and, even in ye olde hardware corner shops burgeoning with long-forgotten items of particular usefulness in small villages dotted the length and breadth of Somerset, the wick of such dimensions was no longer available. She figured she just might find one in India and brought the nozzle to the desk.

Peacock was on duty: *You can get this at the market,* he said. The thought of jostling markets was too daunting for Anna to contemplate. She appealed to chivalry and was rewarded with, later that afternoon, a ball of poorly spun orange string. *You use double,* said Peacock, adding, to silence her obvious questionmark, *you cannot be buying this stuff,* meaning the correct wicks, *in Delhi, we are an upmarket society; this you will find in the,* sniff, *poor villages. We are not using such things here.*

There, she was firmly put in her place – a poor villager! She grinned, knowing her position in his scheme of things.

No, she wouldn't share lunch with Arun *there*; there, eyes were censoring, critical, she was a 'phoren' woman, lunching with a taxi driver ... Anna quelled the thought, besides, it was just possible, as her Hindi was non-existent, that Arun's English didn't appear to extend to much more; he did not initiate sustained conversation, lunch may have strained the invisible that linked them. He understood her though, they had passed a coconut wallah on the kerb and she said, more to herself in memory of Malaya, *I'd love some coconut water*, and he stopped his car on the head of a pin and called out for one to be brought to her window – *and make it the best* – meant he understood her very well indeed.

Now, nearly old, she had been drawn back to the land whose quiescent wisdom had informed everything real she had done in her life, and informed every great teacher she had known. A quarter of a century had passed since she had sat with Father Bede in his hut at Shantivanam; Mrs Tweedie had died twenty years ago; over forty years had passed since Paramahansa Satyananda had confirmed her name, Sita Devi. Her revelation of Sita and the golden thread which had bound her, guided her, to that mythical forest just a week ago sent ripples into her very cells, pulses of Far Memory.

Anna's reverie, heightened by the moment's sifting of memories, spun backward. *Anna*, who was that small child she once knew in the reflection of the quiet splendour of Anastasia? Those mountains, whose grandeur wrung tears from her heart, would she ever see them again? Haimavati, the Daughter of the Snow, the essential feminine ... Deep in her reveries a voice rose softly, she keened her inner ear, eavesdropped on the Heart of hearts: *Eves*-dropped; she re-defined the Forever Feminine. She heard Moti's voice from the distant past, Moti and his wife shadowed in his tiny shop as Anna had looked up to her first glimpse of the Himalayas:

We adore the Divine Mother without asking why and we love God unreasonably, she heard the soft echo of the old man's voice in the hollow of her heart, *our very lives exist from the Existence of God.*

After her experience in the Forest with the Heavenly Horse Anna had begun to understand her beloved aunt Anastasia in a different way; and Anastasia had shared much with Anna over their all too short years together.

Shortly before she died Anastasia had asked her to bring their tiffin on to the small veranda, it was Tuesday and Nandini was away. The day was chill, Anastasia brought thick shawls with her, sat facing the snows; she was quiet, everything around her was quiet, faint rustles in the deodars told Anna even the birds had stilled. Devadasi was sitting, reluctantly in the chill but needing to be close, she would jump on Anastasia's lap any minute. In this grave and measured moment Anna sensed would be one-of-*those*-moments. Her skin prickled.

Anastasia began slowly, telling her how the Divine Power finds its way into the heart and mind and body and blood through each opening of love – *and loss*. The blood then beats to the rhythm of Creation, the whole body sounds the Call until the heart utters the Holy

Name, and the mind bends the will to Obey the Law: *Something happens in that Obedience,* Anastasia had said quietly, *something occurs that is beyond death and beyond the world; everything is contained in that Divine Force of Shakti. For Indians, for Hindus, that same Force is resident in every temple, every shrine, every leaf, every dust mote. When a Hindu places her forehead on the stone step of even what might appear to us as the crudest shrine, her action is a living symbol. Discover that same fixed point in yourself. Discover the axis of life around which circulate all the actions of your inner life. Then you will live in that secret power and you will never speak of it, for it will be you. You will be unable to observe any separation between you and the Power of your Will to Obey – yet you must keep awake the Sakshi Bodha, the Silent Witness within. There is the paradox.* She paused, poured another chai, warm and heavily spiced with fresh gingerroot, cardamom, coriander, cloves with only a little sugar.

It will not be easy for you, she continued, and Anna would remember forever the blue grey gaze, *you have layers of covering, everyone has layers of coverings, but yours will be wrenched from you as loss upon loss lays access to your soul. You have not chosen a simple Path. But, Anna my dear, always remember, Somewhere you said Yes! Yes, to Everything.*

Her hours in the quiet embrace of the romantic garden were the perfect codicil to end her long pilgrimage but now the witching hour was approaching, too early after an earlier breakfast, and Anna needed to return to the hotel. After lunch she would be alone, to rest, and to pack, saying farewell prayers in gratitude for this journey of journeys, quietly, in her room between lunch and siesta. Arun would return some hours later to take her to the airport for her four a.m. flight.

Just on midnight she walked to Reception with her box of survival rations from the restaurant kitchen, to cover the flight's irregular meal times, and waited, tensely, until a call, and: *Madam your driver waits downstairs.* More goodbyes, thank you's and Anna wheeled her tiny bag to the lift and down to his waiting car.

Arun welcomed her. Since he had dropped her in the morning he had been driving, taking fares all over the city, and he now looked drawn with tiredness; it was midnight. They drove with little conversation to the airport and once there Anna was momentarily nonplussed, couldn't gather her bags, garner her whereabouts, the tension between them acute, palpable. She managed to get out of the taxi; Arun got out too and lifted her bag, passed it to her. They stood for an eternity of a moment unable to speak, until he returned to his cab – and sat. Anna stood, still momentarily lost, waving him to go, indicating above the din of cars coming and going that she would wait until he was out of sight. Inside the cab Arun shook his head, *no*, she could see him saying, waving, *you go, to the entrance*. She shook her head. He got out of the car and oblivious to all he walked towards her, a magic and lavender space unpinned from the hours and the world between them, suspended between realities, as horns beeped and the snappily efficient traffic conductor shouted for him to move his taxi. Chaos ensued, but: *I cannot be happy,* he said, *until I see you are safe at the Terminal doors* – they were across the forecourt – *only I will leave when you are there.*

Oh, Arun, Anna sighed, *then please make a selfie of us on your phone and email it to me.* He did, she smiled and obediently walked across to the Terminal doors, turning to wave him a final goodbye. As he drove away, slowly, she felt something being torn away, something pull her with him. She turned and went inside.

The flight home was forgettable, as flights in cattle class mostly are. At Heathrow Anna took the Piccadilly line to Hammersmith, and, with five hours to wait before the Berry coach to Glastonbury, boarded a 391 London bus to Kew Gardens, lunching at the Coach and Horses, leaving her bag with them, the staff knew her, a gift from her Secret Agent in January occasioned that, to spend the warm Spring afternoon in the gardens, wishing she could show Arun *her* paradise. High in the trees nesting pairs of Indian ring-necked parakeets, shimmering green escapees, gave doubt to her homecoming. She took refuge with Marianne North, claimed a seat and remained an hour in front of the Sacred Plants of India. India had reclaimed her heart and her soul.

Home. Anna thanked the Presences protecting her tiny sanctuary of a home, the same home which had been the cause of such suffering for years, but which neighbouring council bungalows were now peopled with reasonableness, a secure gate she had endlessly petitioned for now putting a stop to wandering drug addicts and the horrors of those earlier, sleepless, years of broken windows, terrifying raids, gangs of giant drug-stupid youths inches from her low bedroom window as they drunkenly, violently stumbled through the arched exit to Magdalene Street. It was still hard to re-adjust, fear still clutched her stomach when she came home from being away, even for a day, what would she find? But her sleep, after long years of fear her sleep told her the awfulness had passed, for these nights she slept soundly. Other residents had begged to be moved, and were, but Anna missed out on that possibility when the rules permitting change changed, she had to remain there, fighting for the protection she knew the holy ground annexed from the Abbey required of her. And it was holy ground, occasioning rare sightings of Benedictine monks passing by her window; she could see them from the corner of her eye, turning her full gaze on them would lose the momentary vision and the spectral monks would fade back into the mists. Anna felt their presence and more besides, every image and ornament in her tiny hermitage held a numinous essence, a story; even her books breathed out a friendliness from their shelves as they jostled for spaces.

She woke early, opened her emails. Her heart leapt at the one name amongst so many: *Arun*. Attached was their selfie, and a simple, dear, message: *When you go I was so sad. Our days together were special. I really miss you n I tell my wife about you. At the airport I do not want to leave you. Now you will be home and I am happy you are safe. You are a wonderful woman. Arun.*

She sat, a long long time, reading his words over and over. This young man had opened her heart and the whole of India had fallen into its boundless capacity. *The worst of going to India,* she said later to a friend who understood these things, *is that afterwards everywhere else, even home, is a kind of exile.* She looked up from her computer – her eyes would search for his name in its inbox each morning – in vain of course, and – laughed. There would only be one email; *could* only be one email. Their worlds had no common boundary, some thresholds between departures too wide to be taken in one stride. She would miss him, miss the invisible magic, miss her space in his heart and for a time would make a Gregorian chant out of his name as she lay down to sleep, knowing that always in between *true* and *false* is ... *maybe*.

On her second evening back home the BBC aired an interview with Judi Dench and Ali Fazal; *Victoria and Abdul* being the film of the moment. *My*, smiled Anna to herself, seeing how the pair sizzled as they sat on the studio sofa, eyes twinkling in the pure enjoyment of each other's company, age irrelevant: *Well he's such a beautiful young man,* grinned the great Dame as Ali chortled adoringly: *And she is such a flirt!* in response to the interviewer's very noticeable gooseberry moment.

Anna remembered Arun, his pure kindness, his *knowing*. His care had warmed the cockles of her heart.

In the romantic garden...

3

A Lavender Codicil

Lavender: a herb which, when pressed, exudes an intense fragrance and on release lingers on the fingers as a memory; traditionally associated with old ladies and old lace.

Anna knew she would not see Arun again. She knew she was unlikely to travel to India again. She had seen her Tiger, had witnessed for Ustad, for all tigers, all the world's creatures really. But ... how the gods smiled as they planned her lavender codicil ...

Every new experience filtered down through Anna's consciousness, rested in depths, added incrementally to a courage she didn't know she had. Out of courage came confidence, and since her return from India after so many decades Anna's confidence sat very close to the surface.

There was an ancient Copper Beech tree in the Abbey grounds, it shone from its own light, so loved, admired, hugged and sung to by multitudes of visitors was it. One late afternoon, in the summer after India, the tiger, and the taxi driver, Anna was drawn to its broad, coruscated trunk and asked her friend accompanying her to take photos leaning against the huge tree's gnarled trunk. The image of Anna's dress, dove-dappled, dappled by beech-light, impressed her mind: *A far cry from dhoti to dotage,* she laughed to Chelça, recalling the dhotis she'd worn in her younger years when she ought to have been beautiful but was too damaged by her mother's mantra to be so. She had come far. She could wear peahen brown with aplomb now, it was a quality, something *other* than just style. She felt it in her womanly wonderfulness. India *had* quickened her soul, her very Self.

Something had shifted after India, something profound. She had barely unpacked the one small cabin case she had taken to India when ...

Dappled by Cooper Beech

4

The Devi Calls – on WhatsApp

Anna rose from her computer, a secondhand notebook, a model known as Yoga with a deep *geru* coloured cover, dreamily appropriate; she named it *Saffron*. She always named inanimate working things like cars and computers, knew naming *empowered* them, created a force field between the implement and the human.

    It was time to prepare breakfast, a simple affair of whatever protein she could find in her fridge. She was not a savvy food shopper, partly emotional, her mother's mantras of not being worth feeding and partly economical, her miniscule budget leant towards a pot for pennies and pounds for travelling; meals, though nutritionally good, fell far short of balanced or bountiful. Breakfast, settled on a small Florentine burnt orange and gold tray on her lap, was eaten in tandem with Breakfast on BBC Radio Three, its classical music usually calming.

    Anna scanned her thoughts. This beautiful fragile planet, which she knew quite well, the turquoise seas and the coral reefs where she once lived, the rainforests and the red desert earth, always called to her. Here, on the other side of the world, she looked out from her window to see the dead leaves of the late spring daffodils giving nourishment to emerging beetles; the Cécile Brunner rose, that exquisitely fragrant *sweetheart rose* so loved all over

France for that past century and a half, now polka dotted with tiny buds bursting with life. All was holy.

August came. Anna had been writing her mosaic of love and loss, The Tiger and the Taxi Driver, its title and its progress catapulted by the intensity of meeting Arun, for six months. Arun was never far from her thoughts, the photograph she had taken in the romantic garden by the purple bougainvillea in March on her last day in India sat on her windowsill, another, tiny, on the faux fireplace. She wondered how to end the book of her life. The thread that linked her with the taxi driver, gossamer fine, had no finality in those wonderings. He seemed ever-present, tangible, on a parallel plane of reality.

In August, the 15th to be exact, a friend, perplexed by Anna's cell-phone phobia, urged her in a no-nonsense way to activate the mobile phone she had been given, under curious circumstances, eighteen months before. Cell-phone-phobic-Anna, not wishing to be reminded of the strange memories surrounding how she had come by a phone she had no use for, had stowed it in the glory-hole. Deborah would go with Anna to Carphone Warehouse to translate phone-jargon and encourage her to get on to WhatsApp, the open sesame to the Indian world. *Oh Grandmother*, said Mohammed and all the tuk-tuk drivers of Anna's acquaintance back in March: *You must not be travelling in India alone without a mobile phone!*

Deborah was more than her word. Anna retrieved the phone from its box and the two women went to the phone shop where, to Anna's delight and budget, *Virgin* was offering a monthly sim contract at a very affordable price. Anna spent the evening gathering up loose bits of paper and various address books and inputting phone numbers. She turned off the phone, put it to charge and went to bed.

    The following morning Anna switched on her new toy. It played an endless melody of triple notes and: *Hey, I'm on WhatsApp,* appeared under countless numbers. This might be fun she thought, as she registered names and – *Arun!* No pre-selected message from him, his name sat above a row of roses and bouquets. Anna's heart, spring-loaded at the sight of his name, flipped over. She held the magical device in her hand, wondering, still, excited, disbelieving; went to get breakfast, returned and the message was still there. She paused long enough to figure out how to respond and then – their communication began. India was far from her mind but Arun's morning messages with little photos from the WhatsApp store lightened her heart; they had little to say to each other but the link was made.

One evening Anna was watching Alexander Armstrong quiz a contestant about her birthplace in India. Anna's ears pricked, though she didn't catch which town in the Kumaon Hills the woman had been born, she heard her say she had recently returned from a Literary Festival. That magic word, *Kumaon.* Anna's head buzzed as Arfie, top dog of *www.dogpile.com*, searched for Literary Festivals in India past and future and came up with *Nainital, October,* just about six weeks away. Nainital, the Queen of Hill Stations in – the Kumaon Hills.

    For three days she pondered, and on the third day WhatsApp'd Arun with: *How do I get to Nainital from Delhi?*

He responded within the hour: *Train is not good idea,* mindful of her need to eat regularly, *it is six hours and train only will go to Haldwani and next 60 to 70 km travel from Texi. If you wish to go with me then I will go with you.*

Anna sat with this breathtaking invitation, waiting, as her head and heart re-connected, before a huge *Yes!* gave her the obedience she wanted: *Yes!* She sent back, and emboldened by dreams asked if it would be possible that he would wait out the duration of the Literary Festival to take her onwards to Kausani, to Landour, to other places she had dreamed of. And the costs?

Yes it is possible and the cost is 15 Rs per km and 250 Rs allowance per day.

A godsend.

She'd write a Will, just as she had when she walked the Camino, death was always round the corner for Anna, was it her Benedictine inheritance from other lives: *to keep death daily before your eyes?* she felt so disconnected from this world. If she died, as many did, on those mountainous and treacherous Indian roads she would bequeath to Arun all that was in her accessible bank account. It was a lot, for her, and would be for him too. And should they both die in those all-too-common accidents on Indian roads, well, it would go to his family – and to pay for a splendid Hindu burial on the Ghats of Varanasi for each of them; going up in flames seemed a wondrous way to leave no footprints: a recognition of the Hindu rightness about death, no taking up earth-space and no memorials for mourning. Anna had noted that there were no graveyards for the million Hindu soldiers seconded to the first, or second, World Wars – *like the*

Aboriginal people, thought Anna, *who won't speak the name of their dead.* A woman from one of the northern clans had told her: *We don't hold our people back when they die, if we say their name they hear us and then they hang about, most of us don't keep photos; we allow their spirit to move on.*

How clean that thought, Anna thought. The young solicitor temping in the Abbey Tea Rooms witnessed the Will for her, there were mild discrepancies in her wording, he comforted her: *With a niche name like yours no one's going to mistake your intentions!*

For now the miniscule costs Arun had proposed prefaced her dream of possibilities, a road trip round the Himalayas would be the gift she would give herself now she was seventy-two.

Reflections: Anna felt her 'work' in Glastonbury was over. She had done some things and initiated other things; touched and been touched by many lives in her Obedience to live at a slant to the Universe. She had forged a courage from adversities that might fell lesser hearts (as indeed they felled hers on many a past occasion); she could look back and know *that* assessment of herself as courageous accurate to a degree. She was equally adamant that Catholicism's true value was in honouring Mary, for Mary's humanity as well as a palimpsest for the Creatrix. Mary had given birth, in her own immaculate way according to the church, to the Lamb of God. Stretching her credulity further Anna recognised the Lamb as symbolic of sacrifice. However, it was Mary who interested her; the boys of the world and of every religious persuasion had had their own way for too long and had perpetrated far too much damage and desecration on this beautiful Earth to command much respect from Anna. Mary was a young a woman, a very young woman if stories were to be believed, and she witnessed the most horrifying death a woman could ever witness of her child, her son. That, Anna knew, was the bottom line, and, for millions of women that should be enough to link them. Yet all Christianities except the Catholic and the Orthodox had airbrushed Mary from their theologies, out of their his-stories and out of their art.

Virgo, she learned from Gary Wiseman, a wise man from Melbourne, who taught her these things many decades ago, was seen in astrology as the potential bridge to the transpersonal world and the natural processes, processes which are also the mysteries of transformation: birth, love, and death are all, in esoteric astrology, the domains of Virgo. Early Catholics, whose cultural roots still recognised the coursing of the stars and the meaning of their constellations, gave the 8th of September, midway through Virgo, to Our Lady for Her 'birthday'. In esoteric symbolism Virgo was the oldest of the Zodiacal signs. Lilith, Isis, Eve and Mary portray the Mother of the World and the Mother of the World will always give birth to a hidden Something.

The Mother of the World is fast fading from human consciousness except in India. Humanity is no longer in awe of the Spirit charged within Matter. *Adab*, respect, respect for just about anything or anyone has all but vanished from much of the western world. Anna found it in France and in Italy, still strongly Catholic countries and, though church attendance may be nominal, the manners and the mores of respect survive. She rued small losses, telling in their simplicity: the stone plaque set in place in 1954 in the wall by the door of the domed French Catholic church in London's Leicester Square had recently vanished. Dom Robert's words: *Tapistry: The Blessed Virgin – Queen of Creation Designed by Dom. Robert O.S.B. and made at Aubusson,* confirming his own great epiphany, had been removed since her first visit there. Vanished! Cut out of the wall! What a to-do. Anna had photographed it; it *had* existed.

Anna felt a genuine sorrow at the removal of the plaque, knew it symbolized a cavalier repudiation of Dom Robert's personal epiphany. But the greatest tragedy was the loss in the West of that glimmer of Light which pointed to: *She, the Creatrix, with God before all ages, before Time itself.* Extinguished, no longer honoured so, Anna rued the loss of that modest stone plaque which in its simplicity held the Cosmic Truth of the *Womb* of Creation. Aditi and Mary, both almost beyond conscious recall, each diminished from the Cosmic Equation. At the end of his life Dom Robert *knew: Nature never deceived, Nature is the true Reality, Nature is the Face of God*, he would say. He would not have access to the word Shakti in the time of his Catholicism, and the mighty word Goddess would not have entered his lexicon, but he *knew* the vivifying force that gave birth to Creation through the Void. Van Gogh, who didn't know Mary at all, echoed Dom Robert's epiphany: *Nature is God, I want to show people who can't see, God.*

With such thoughts fully charged Anna was more than ready to return to India, to the land of the Devi, to Shakti, and to the Goddess. She chose 29th September to leave;

the Day dedicated to Archangels and the anniversary of her unknown grandmother's birth in Paris. Only three people knew of Anna's proposed trip. One friend emailed asking if she had a reply from her 'cabbie': *I expect you are having butterflies,* she wrote, *it seems a very brave and adventurous thing to do. But you did it a little while ago and then you didn't have a Magic Taxi nor indeed a potential Swain to look forward to (and worry about you). The Things You Do. I'm so proud ... xjx.* Anna responded happily that her swashbuckling swain had WhatsApp'd telling her to: *Relex and I will be sur of your safety.* She would be in good hands.

Yet, it was a brave thing to do – the pair wouldn't be able to communicate with any real dialogue or mutual language, Arun's English was endearing and his text to her: *Do you speak Hindi?* made her chuckle. *Not a known language in Somerset,* she replied. Arun-her-Hero relied on pronouncement and presumptions: *We will do this, do that, come, come, come* ... while she choreographed the Vision and the Mapping. He rose to each new place, some new for each of them, with curiosity and engagement. But: *No,* he would wave away the waiter offering a choice of naan, *we will have butter-roti* ... Anna laughed at his puzzlement when she murmured she just might like plain naan. She ate the butter-roti. She put all her dreams into his hands and extended her wish list exponentially. They would make a Road Trip around the Kumaon Hills and beyond, in the Shadow of the Devi, a codicil adventure. In her own way she was weaving a web of threads of the Divine Feminine from Our Lady's Dowry of England through the ancient warp of the Goddess, far away from English consciousness, to the Devi Who would guide their pilgrimage.

The day before she left Glastonbury Anna took Deborah into St Mary's. A few days beforehand three icons of great significance had been blessed in situ, Anna wanted to show them to her friend for it was Anna who had been responsible for their being there.

Years before, Anna had twinned Glastonbury with Patmos in Perpetuity, the only Twinning in Perpetuity in Britain, as befits the incorruptibility of Saint Joseph and Saint John, the titular saints who held each place as his own. She was quietly proud of her Obedience to the Imperative she had been given in the Cave of the Apocalypse on Patmos in 2007 and the encouragement she had received from Abbot Antipas to return to Glastonbury with his blessing to twin the two holy islands. The two saints knew the living Christ, and the Greeks knew the story of Joseph who came to Glastonbury with his holy staff of greenwood to plant in the soil where it took root, as greenwood does, and grew into the legendary Holy Thorn: *More than most English know, or even care about knowing*, thought Anna ruefully.

The Twinning in 2009 proved a splendid four day event and Anna would smile and say afterwards: *The greatest thing to happen to Glastonbury since Queen Anne gave the town two Royal Mace in 1702.* Some years later a young woman from Athens made contact with her through the Patmos link. Katerina Karoussos had a story to tell.

Her father had dreamed of bridging the Schism, that needling sore between the Orthodox and Catholic churches for more than a thousand years, through his art and his spirituality. He was an artist of the highest calling, icons and frescos being his inspiration and his spiritual discipline. His first commission was St Panteleimon in Athens, the commission for St Andrew in Patras was his too. When Pope John Paul ll gave the Church of St Teodoro on Palatine Hill to the Ecumenical Patriarchate of Constantinople, Iannis Karoussos was chosen to paint its Dome and Temple. His work restoring the Church of St Panteleimon in Athens took twenty-five years; he died before it was quite complete, his devoted students finished his work.

Abiding by her father's Vision, and aware that his love for the Holy Island of Patmos linked directly with Glastonbury, Katerina approached Anna, as the coordinator and inspiratrix of their Twinning, with the offer of three of her father's most remarkable icons to the town. Where else but the Catholic Church would an icon of Holy Mary be welcomed and considered sacred? Certainly not in the mediaeval once-Catholic church of St John which had not venerated a statue of Mary for five hundred years.

Anna made an executive decision within a nano-second of meeting Katerina — St Mary's was the only hallowed sanctuary in Glastonbury that would honour such a gift in Perpetuity and on a daily basis. *Come*, she said, and the two women ran down Magdalene Street to St Mary's just in time to see Father James heading off for a funeral. They begged him for five minutes. He was more into bungee jumping than icons but agreed to their breathless proposal. Two years later almost to the day, it came to pass. Father James had moved on and another priest had taken his place. His recognition of the gravitas of Katerina's offer, and her father's intention, inspired the new incumbent to have them set in the Chapel under the stained glass window of the Virgin and Child in a simple retable of Somerset oak, the whole rimmed with rare Somerset bog oak, petrified in the peat fields of the Levels for over six thousand years and as hard as rock, as imperishable as Perpetuity, the completed Triptych was magnificent.

On the night they were blessed by Archbishop Gregorios of Thyateira and Great Britain, Anna stood by the spirit screen of icons and felt a longing so swift and unbidden it took her breath away. Katerina had not bequeathed Christ the Pantocrator but Christ the Life Giver. The word Zoé, the name Anna had created as a comfortable nom-de-plume when Australians failed with Nigelle, was written on Christ's right, ζωή - Life. It would remain in perpetuity when she was long forgotten; an essence of her *being* would remain. Tears filled her eyes with holy and humble pride that she had been Obedient.

Later, as the Archbishop moved away from the icons after blessing them a gasp went up from the congregation – light flowed through the roundel window with such intensity as to render speech redundant. All that could be heard were cameras and phones clicking. Light could not flow upwards, this did, and besides, that window was blocked from sunlight by Morrisons, the supermarket behind it. It was a holy moment.

Kim, the lynchpin of the parish, came to speak to the two women. As they were leaving Anna mentioned she would not be around for a few weeks, and: *Where are you going?* asked Kim. *Off on a Himalayan road trip*, Anna replied. Kim's face lit up: *Whereabouts?* she asked and

Anna's mention of Nainital went no further. *Nainital!* exclaimed Kim, *my father went to school there, St Joseph's! None of the family has ever been!* It was one of those moments of Timing again. Something would unfold.

The next morning Anna left her prayers with the Presences in her tiny home, pulled the curtains closed, checked all that needed to be off was off, stepped outside to a cool dry (thank You) daybreak and wheeled her old carryon to the bus stop. Passing her Indian friend from Gujarat, by way of Southall, now owner of the Convenience store in Magdalene Street, she popped her head in to say cheerio. He gave her a hug and asked God to keep her safe and, as she stood waiting for the Berry Bus, Marion came to say goodbye and to wait with her until the bus came. Marion who so loved India and who had known Mrs Tweedie, knew what Anna's pilgrimage to Kausani meant. Just as she was boarding, Arun WhatsApp'd a message to: *Be safe, I will be here for you.* She took photos of Stonehenge as the coach passed and sent them to Deborah, to Arun – she was on her way!

The Creatrix of Dom Robert

The Holy Light

5

Ossam, just Ossam ...

Anna came out of the airport, began to walk across the concourse, tired from the flight and momentarily nonplussed at not seeing Arun. He had promised to be there, to take her for ossam coffee and breakfast. Reading *ossam* on his WhatsApp message a few days beforehand had confused her mightily. Allergic to google's world domination she sent Arfie of *dogpile.com* to look up cafés in Delhi but he could find none named Ossam. She took the word for a walk to Street, the neighbouring village to Glastonbury, two miles distant, sounding *ossam* the whole way. After a turmeric latte from Prêt à Manger she began the walk home suddenly struck by a Eureka Moment: *Ossam! Awesome! Of course!* Where else would her awesome young man offer to take her but a café with *ossam* coffee?

But where was he now? She wondered, fleetingly, if she would even recognize him, she only had the snapshot of that bougainvillea moment and a blurred midnight selfie. Behind the waiting group at the exit doors a swift movement caught her eye, and then the broadest smile, and suddenly they were giving each other the warmest welcome and tentative hugs – did a young Hindu male accept hugs from an old foreign woman? Anna hesitated. Arun had already taken her carryon, amazed that this was her total luggage: *And I have eighteen changes of clothes in there,* she had told astonished friends, *a scarf to ring every change, dress up, dress down, I'll face any occasion.*

They reached his familiar old car, each chattering happily and nervously to the other, he held open the door for her and on the passenger seat – a garden-sized bouquet of red roses and blue orchids greeted her. Settled in the car he asked for her phone, opened the back, unclipped her sim card, folded it into a tiny piece of newspaper in precise folds which, unfolded, had revealed an Indian sim card which he deftly reinserted into her phone. Done in seconds, he smiled into her eyes and her world dissolved. Anna knew her Adventure was going to be Extraordinary.

We will leave on Gandhiji's birthday, said Arun, *it is a national holiday, Delhi will be all-closed, good time to leave.* Anna had a *frisson* of doubt as to that wisdom; holidays in most countries she knew meant the hordes set to the roads. She had brought with her a tiny red embossed leather diptych, containing two diminutive images, one of her own French Madonna with the pearls, and the other of Ma Durga on her tiger; these were for their dashboard, Travelling Protectors. Arun smiled to himself, this woman was alright, he was right to say yes to an adventure with her. He was charmed.

They spent the following day doing simple sight-seeing, not to anywhere Anna particularly wanted to go, although passing the Sacred Heart they stopped and went in for a few minutes. Arun declared: *Our journey begins now.* The B & B in Defence Colony which Anna had booked via email was, frankly, a spotlessly clean disaster. The owner wrote excellent English, had created an enticing story of home-made jams and home-baked bread but forbore from saying it was served stale, from frozen, resulting in a slap of pap. Real choice of real food was limited in the extreme. The menu, well written, did not mention there was no chef, resident or otherwise, to live up to the story. The host did not live on the premises either and *real* food a long convoluted walk to dubious eating places. The choice of tea, much lauded on paper, came as *unwrapped* teabags, open to the air for how long? to taste as stale as grave dust. Anna's misjudgement of the place as she read it, was total. Why oh why hadn't she gone back to Palace Heights? Arun was elastic in his promised arrival times which added to the bodily stress of needing to time her eating. Carrying snacks was a non-event; there was no fridge in her room, no shops nearby.

On the morning of their proposed departure Arun phoned to duplicate her alarm at 4am. He would be there at 5 a.m. with an omelette, cooked even earlier by his wife, he assured her, and they would start their journey.

Having no resident staff, her 'packed breakfast' from the evening before of two sodden pieces of stale toast and its square of yellow processed pap, was unworthy to feed to street dogs. Arun was late. Arun was so late. At his request the previous evening Anna had handed over a total of £400. Nearly forty thousand rupees, many times more than his request by WhatsApp. It was all the money she had brought with her to pay for their accommodations. He calculated they would be travelling 3,500 kilometres. Anna, forty years Australia-wise when it came to travelling long distances, blinked at his guesstimated pre-calculations; at most it would be fifteen hundred kilometres. She would set the car's odometer when they started the journey.

Time passed. She made full use of her WhatsApp messaging but received nothing in reply. Alarm shadowed her sanguine, the mobile phone was a body part for Arun, for all Indians, why no response? Awake, her body needed ballast, there was nothing available, no staff to provide it, and time, in hours, ticked by. Anna lay on the large sofa in the meeting area breathing in trust along with her mantra. Still jet-lagged, these extra hours of deep sleep would have been welcome. Daylight broke through the long French windows, but still no Arun. If she lay very still her body would use up less energy, perhaps she could last until breakfast when the staff would arrive. As for reaching Nainital, she would gird up her loins, eat well later, pack picnics and travel up by train tomorrow. She had paid her accommodation there in advance.

Calmed slightly by her reassessment of her situation Anna remained supine on the sofa. Three hours passed. The door opened, a surprised houseboy came in, not expecting her to be there. Anna felt a cloud of anger at Arun, he had badly let her down. She asked for tea. The door opened again, Arun's smiling face... by now hers had a thunderous cast. *Sorry, sorry, sorry,* he said, *come, come, come I will explain,* as he picked up her carryon and ran downstairs. Anna, dazed without food, followed more hesitantly. But where was his familiar car? *This one:* he said, putting her bag in the boot. She sat herself in the front, and a stranger sat in the driver's seat. *What's going on:* she snapped as Arun got in the back, *I always sit in the front and I want to sit next to you so we can talk – who is this man?*

It was not the most auspicious of beginnings. Tired, hungry, sugar-melt imminent, coma round the corner, Anna was not coping. She needed control over her food timing, like the medicine it was, and hero had let her down hugely. They should, by now, have been a hundred kilometres out of Delhi and eating a substantial breakfast at a good roadside durbar. She got out of the front and into the back where Arun was now sitting like a lordling, issuing instructions to the unknown driver. Born under a mutable star Anna could turn disaster round on the head of a pin when change presented itself as *fait accompli*. She unpacked the lukewarm but now greasy omelette carefully cooked by his wife hours beforehand, began to eat a little, felt her shaking body respond, and re-considered the benefits of sitting next to her godling for the journey, able to engage in undistracted conversation while the driver, drove. *Maybe for the best*, she said to Self, as her anger dissolving in the balm of Arun's charm.

Farmers were rioting at the exit from Delhi, massed road blocks of armed police were directing all traffic away from access, re-direction meant further unconscionable delays and another hefty amount of State taxes to be paid to exit from a different route. Anna requested explanation, she would, after all, be paying and she had already handed Arun all the cash she had brought with her. That four hundred pounds was all she had until she could access an ATM. The sum was a far cry from his original lure and would rise ever more exponentially as she would slowly discover. But, for now, she just wanted explanations for the car, the driver, and Arun's tardiness.

The sun was rising above the treeline, late for the chanting of the Gayatri mantra, but before answering her reasonable request both Arun and the driver began to chant, Arun with his palms together, facing right, towards the sun. Anna joined in. It was a mantra she knew and loved. Chanting it this way, in unison, filled her heart and another part of her realised her head had calmed completely as the ancient prayer enfolded her in the new day. The driver commented to Arun that she sang Sanskrit perfectly, unlike most westerners who used unsuitable English intonation and incorrect syllable stress. Arun translated his Hindi to her. *A little flattery...* mused Anna, and in her new-found equilibrium asked gently again for true explanations.

I am so sorry, said Arun and his voice changed timbre, his expressive eyes held hers, Anna inwardly sighed, she knew he could charm the gold from his grandmother's teeth: *This morning I go to my car at four o'clock ready to come to pick you up and the back window has been broken in. I have to wake my travel agent and he has to wake a driver and we have to wait for a car and I have hired this man for the whole journey and I have paid, you will not pay ...*

Of course she would pay, it was costed in to the triple distance he'd quoted in his guesstimate the previous evening even before the broken window story. But at that moment Anna hadn't registered the plot, if plot it was. Instead she recalled a dream she'd had at three o'clock that morning, which had shocked her into waking to look at her clock before the four o'clock alarm:

Kali Ma visited me this morning at three o'clock. She has never visited me before. I took note of the time and her presence. She was calm, but she waved one of her raised hands, the one holding the sword. I knew it was a warning of something but I wasn't afraid because she smiled. I am always comforted by the image of Kali, she uses her sword to protect the truth found behind the act. I happen to have been given a Kali bija mantra forty years ago when I was initiated as a swami but I found it was too powerful then to use. Perhaps I should begin! She has obviously been looking after me all this time.

Arun watched her face intently as Anna revealed this unknown history of herself. He took it seriously: *Kali Ma came to you this morning? At three o'clock? Then that was the time my car was broken into. We must, you must, make offering to Ma Ganga as we cross. Here,* he said, opening a small bag, *your Ma's are safe, I take them inside my house last night,* and he positioned her travelling diptych on the top of the back seat.

Long slow kilometres further on, the whole of Delhi appeared to be on the move for Gandhi's birthday, Arun called the driver to halt on the highway crossing the great river while he beckoned Anna from the car, holding the door open for her. He gave her a coin, she knew to cast it wide, it would fall deep, and with it offer her prayer. He photographed her in the act, a treasured moment. Back in the car he showed her the photograph, remarking on the singular synchronicity of colour: Anna was wearing a long, deep charcoal grey linen dress and a golden ochre open-weave cardigan; Ma Ganga was a deep charcoal grey and the stone parapet Anna was leaning over painted golden ochre. The match was exquisite. Arun smiled, he knew the journey with this woman would be – Extraordinary.

The drive of six, or was it seven, hours swinging around all the torturous devil's elbows of the road into the mountain foothills of the Kumaon Hills was best forgotten as Anna grew fainter and greener by the hour. No such things as toilets existed along the snaking road, and when they did stop at waterfalls and a makeshift café Anna availed herself of a shrub further up the hill rather than lock herself into the single stinking unplumbed unisex tin cubicle nominally referred to as 'W.C.' As for the scenic waterfalls – thick with discarded plastic water bottles and all manner of human detritus – they were far below her radar for scenic attraction. She knew she sounded surly, but by then, she was. And being Anna, unable to pretend otherwise. The food at the pit stop was poor and she was neither consoled nor her body balanced. Yes, it would be a difficult trip despite all assurances to '*Relex*'.

Yet – rounding the final bend, after endless signposts with many kilometres still announcing the day's journey's end, Anna's heart soared at the sight of the jade green lake beyond and below the high forested hills that earned Nainital the title of Queen of Hill Stations. The driver stopped, the Y junction revealed a hotel of colonial proportions on the hill to the left, he swung an uphill U – and there they parked. Arun arranged two good size suites, Mrs Arun would be paying of course, but registering with an Indian national meant she would not be charged the tourist rate of £130 a night, so Arun cheerfully informed her, but only £30 a night for each of their rooms. The driver would sleep in the car. She agreed to everything and, relieved of their bags, the pair walked down the Upper Mall, forbidden to Indians during the Raj, to discover Moti Mahal, with the best of food.

And so to bed.

The morning began with a selfie by himself of them both in the empty dining room waiting for breakfast served piecemeal by waiters, raising inefficiency to an art form, unaware of time and motion. Breakfast was a novel interpretation of the original hot-buttered toast so loved by the English, but butter-tost was a radical departure from toast and butter. A large copper bowl brimming with hot melted butter was brought out – to be spooned on to any temperature toast, its anticipated culinary sensuousness lost in translation. Eventually an omelette arrived and Anna's body and soul reached a certain balance. Strengthened and somewhat satisfied, she could set off with Arun *à deux* to the Naini Devi Temple on the shore close by the hotel's end of the lake.

Arun bought offerings for them each. Set on a tin plate was a rich choice of sparkly trappings for the Goddess as well as little packets of sweets, puffed rice, fruits and incense to thank Her for their safe-coming. As Anna stepped away from her sandals and padded barefoot across the sun-warmed tiles a sense of homecoming so overwhelmed her she sank cross-legged to the ground, her full silk skirt rippling around her. That ancient name sounded again: *Sita Devi*. Arun, Ever-Ready, took the opportunity to savour the moment with a photo.

Reluctantly she rose, followed him to the sanctum sanctorum, lit the incense, offered the gifts, said prayers. As they left the temple, Anna didn't catch the exchange,

Arun held two packets of offerings, one for each of them returned by the pujari. To her delight Anna's contained a strip of dizzying red and gold tinsel edged cloth, four sticks of incense, two deep red and gold glass bangles and the return of a small packet of puffed rice much of which had been fed to the sacred carp of the Lake but what remained she would take home to feed the sacred carp in the pond of Glastonbury Abbey.

After lunch Anna and Arun agreed on an Adventure, they would make it their mission to 'find' Kim's father. Calling the driver, whose name Anna hadn't yet clocked, aware as she was of her resistance to the man whose piercing mobile phone ring shattered every conversation and whose presence intruded on her company with unwanted familiarity, they set off up the opposite mountain to St Joseph's, losing their way at the summit to enter the private estate of some MP before being shown the correct entrance to the prestigious school.

    The chowkidar, employed to prevent stray 'phoren' women from entering an elite Boys College without an appointment and locals from wandering in to enjoy the spectacular views and cavalier parents demanding the attentions of the Principal at their whim, stopped the pair at his sentry box. The Head Master granted an appointment for one hour only. This was not the hour. The chowkidar had little English, Arun rattled off a story in Hindi. Anna had suggested he say it was her father's records she was looking for in an attempt to lend credence to her being there. That didn't work. Chowkidar turned to ask her: *Who? When? Name?* Anna hadn't a clue.

Her giggles at being uncovered in a shameless lie won him over – wagging his head and smiling, he waved them pass.

Next stop, the headmaster. Arun remained respectfully on the splendid steps of the grand old building while Anna went on through the quadrangle, asking a group of charmingly mannered young boys where she could find the Head. In exquisite English they directed her up the stairs, along the corridor – and she tentatively knocked on his door. A small elegant man welcomed her in, she confessed her shameless lie and told the true story and together they guesstimated the age of Kim's father and then guessed at the years of his attendance to arrive at three vast Dickensian ledgers of names, the first of which opened at – Woolmer! Two Woolmers! Brothers! Dr Peter smiled broadly and declared it a small miracle. Unable to resist, Anna grinned: *Well, I do have a hotline to Holy Mary! I share Her birthday!* Photographs secured, the two parted warmly and: *That was a verrry emotional moment*, said Arun in his endearing English, snapping selfies of them both on the grand steps before winding their way down the cobbled driveway to the car, pausing first by the chowkidar at his sentry to show her photograph of the ledger page and to thank him. From the corner of her eye she saw Arun hand him a tip. *How kind*, thought Anna, to whom the gesture hadn't occurred. She so liked this young man whose manners were revealing themselves as impeccable.

Stepping outside that privileged world the two of them were gently hi-jacked by Father Jerome from Saint Francis Home, the Capuchin neighbour to Saint Joseph's. Father Jerome, wreathed in smiles and warmth, led them up the garden path to his simple church, inviting them to attend Mass and the Feast of Saint Francis the following evening. They promised to return.

Their next days were taken up with boating on the lake, a trip in the cable car to the summits overlooking Nainital, suggestions from Arun that she wouldn't have known of without him. Riding little Marwari ponies up to Tiffin Top was another treat he had in store for her, though Anna would have preferred walking, the cobbles treacherous to the hooves even of hill ponies. She consoled herself by knowing her money at least fed them. Walk was a four-letter word for her dark godling, who wouldn't, and who told everyone and anyone that Anna, now in his charge and his responsibility, at the age of sixty-eight had: *Walked 700 kilometres, Christian pilgrimage you know...* when he protested at her suggestion they *walk* up to the spectacular outcrop where Dorothy Kellet once sat to paint away the day sometime in the '30's and Anna had laughed and told him of her Camino. He hadn't believe her, until she showed him a series of photographs stowed on Candy which proved time and place and distances. He was impressed. *Verrry* impressed! He wouldn't consider it even at thirty-five.

6

"Welcome, Distinguished Guests"

Prasada Bhavan, lately renamed Abbotsford, was a true haven and the venue for Himalayan Echoes, the Literary Festival founded by Janhavi Prasada. It drew together distinguished authors, environmentalists, mountaineers, historians, poets, film-makers and artists to share their passions with an audience who, for Anna, would dazzle in the autumn light like a charm of hummingbirds. Mingling and listening to her mother tongue spoken so well, so nuanced, so rich in vocabulary was thrilling for a word person. Anna would say hand on heart that only in India, amongst those who spoke it, were the values of the English language maintained. She heard it used sublimely, spoken correctly, and pronounced endearingly.

Beautiful Janhavi's beautiful mother (*Indian women are so beautiful*, mused Anna) Kanta, welcomed her with a smile and an entire caddy of Lapsang Souchong, brought especially for Anna from Delhi. After that, what could Anna do but love everything and everyone. She had booked to arrive before the Festival, to settle in, re-orient herself to India, feel where she had landed. Listening to Kanta tell her story of restoring her beautiful home, left neglected when her husband died, filled her with admiration for the attention to detail: *Even the rugs, when I unrolled them, were mildewed and moth-eaten, I sent them to Kashmir to a master craftsmen to repair; and the woodwork, the windows, the roof, all were rotten. It cost a crore to restore, but it is beautiful.* How beautiful indeed, agreed Anna, who would have done exactly the same given the circumstances. She warmed to Kanta, sitting there at the tea table with the caddy of Lapsang Souching, the flower-framed lawn falling away to the valley forest beyond. A shimmering silence captured the moment as the late afternoon sun lit like amber the autumn cosmos, the marigolds, the far hills.

Tibetans had begun their exile to India in the 1950's when the wallpaper of ominous signs shrouded their land. His Holiness the Darling Lama delayed his own escape but blessed those who went before him. Nainital had welcomed the refugees for sixty years and banners thanking India and Indians for their safe haven were displayed in the market place by the Devi Temple. Dr Lobsang Sangay, President in Exile, was the special guest of honour at the Literary Festival, invited to give the Opening Address. Anna was immensely moved by his message, his candour and the *frisson* of mild dissension he displayed in conversation with Patrick French, *alma mater* Ampleforth, whose own book *Tibet, Tibet* aired erudite, but personal, opinions. Like Pope Francis, Dr Sangay used his profile as a force of good, for peace, for waking people up to the enormous responsibility we all must share in preserving the planet, recognising humanity's debt not just to itself but to All Beings – to the last blade of grass. Elliptically acknowledging Tibetan gratitude to India's safe haven for HH the Dalai Lama and all Tibetans fleeing the Chinese invasion, Dr Sangay, in one subtle sentence, reminded his distinguished audience that Hindu's most revered lake, Manasarovar, and most revered mountains are in – Tibet.

On a short walk down the steep private driveway to the house the following morning Anna found herself in conversation with a Tibetan photographing the banner high above the drive, spanning it, welcoming Dr Sangay. She posed beneath it for her own photo, and the young man asked if he could interview her, he was Dr Sangay's PA. Caught on the hop, Anna was unprepared, but rose sheroically to the occasion by citing the preferred topics of both spiritual and temporal leaders to use their platforms as Pilgrims for Peace. She felt she had added her bit, naive though she knew her pronouncement to be.

Over the next days Anna was dazzled by the luminaries she heard there on the dias: Stephen Alter – Anna ran to buy the last copy of his *In the Jungles of the Night*; Shobhaa De: *free ourselves from the cage of age*, said this striking self-confessed feminist of seventy-two going on forty-eight; Kavita Khosa (over tea Anna shared with her a few tears as they spoke of the loss of ninety-five percent of India's wild tigers); Harish Kapadia (a day listening to his mountaineering stories would only touch the tip of his 'edge of seat' experiences); Anuradha Roy, *All The Lives We Never Lived;* how Anna could identify with that. She was stimulated and enriched and wished her days here would escalate to years.

As soon as she had activated WhatsApp, had coordinated her codicil adventure, had reserved a place at Himalayan Echoes, had boldly mentioned her preferred tea, Anna had emailed one of the two gentlemen whom she had met at Sitabani in March to invite he and his cousin to join her at the Festival. It was, after all, Jaideep's delayed luncheon appointment in Delhi which now occasioned her attendance at the Festival. Her auspicious meeting with the taxi driver, her obedient leap into the twentieth century activating a mobile phone: with hindsight she found the story so rich with synchronicities she wanted the two men to meet Arun, a kind of happy ever after ending. It was her way of saying: *Look! Isn't life a marvellous thing! If you hadn't been late, if I hadn't had to call a taxi, if that particular driver hadn't come, if he hadn't recognised my shaking state, I would not be here at this wonderful gathering! How little we know of the fates and our destinies!*

They did come, both men, it was a joy to catch up with them. Literary festivals were not on their circuit of shared interests and they loved every minute, appreciated every speaker, were warmed by the hospitality they received, made proud by the Bengal connections. Inviting them was a very right thing to do. Looking for a window of opportunity to introduce them to Arun, who held her life in his hands as the pair would venture into the wilder mountains, was not. The window remained closed, exactly as it should.

As she mingled and sat with other delegates and attendees Anna was asked what brought her to the Festival from so far away. Her reply, to one seriously attractive older woman she had noticed earlier, that reading Felicity Kendall's *White Cargo* as her last choice before leaving opened her heart to all that she remembered of India from decades ago. Wiser now, she kept the story of Arun and WhatsApp to herself. And as well she did, for the name Felicity Kendall brought forth another story: the woman she was speaking with introduced herself as Shashi Kapoor's sister! Jennifer Kendall had married heartthrob Shashi, and his sister babysat the much younger Felicity. Suman told a golden story of Goldie Hawn arriving at her own enterprise, Camp Corbett, under dramatic circumstances. Anna stowed another cameo moment in her mental dillybag.

Meeting these women stirred something in Anna: the life she had not been able to live. A life her privileged family had, and did, live. A veiled sadness hovered in the wings, caught on the threshold of consciousness. She no longer felt betrayed or empty, but how different her life would have been with the education accorded her siblings and half-siblings. Education would have assured her a comfortable path, but would it have unfolded into the Extraordinary? Anna held all life in a crucible of sacredness and, now she was old, her own life too. It had been a long journey to womanly wonderfulness and she was aware that, at times, sacred magic courted her. She had been told it often, and by many, and now she accepted it. This reflected residue of acceptance brought her a strange closeness to Arun, down in the town in another hotel. His heartfelt responses and his insight into the *who* she would come to know rather well during this, her Lavender Codicil, was another kind of magic.

All too soon the Festival was over. The goodbyes pulled at Anna, despite having made only the most fleeting connections, light and elusive and as ephemeral as the mists that shrouded the Hills of Kumaon – or the Isle of Avalon. She WhatsApp'd Arun, asked him to collect her at four o'clock after the final speaker, she would like to stand by the lake, to think, to process the past days riches before the farewell Gala dinner that evening. They planned to leave in the morning at eight, there would be no time then.

He came, the driver took them to the lake, left them together. The aurelian afternoon light reminded Anna of gilded icons, framing the central wholeness in translucent amber, golden, soft, a shimmering cadenza to her days of wonder, shining undiminished despite her sudden Slough of Despond.

Arun watched Anna turn towards the green still water of the lake, lean on the wrought iron rail, head bowed, a glowing nimbus of low sunlight passing through her short fine hair. He felt the weight of her heart, stood close but behind, watchful. She was … surely … *crying*? She turned toward him. The moment spoke of soul on the winds of the centuries, inhabiting a new garment of experiences, awakening memories of deep attachments, spanning galaxies. Quietly he watched her tears. The golden sunlight backlit her as firelight singes angel wings: *Why do you cry?* he asked of this small woman who stood in the background of a painting too fragile for canvas.

Anna barely knew how to explain her forest of feelings, caught and spun in a web they jostled for clarity: a profound regret at having to leave the elite company of minds and hearts gathered for the Festival; the fragility of time granted to she and Arun on this small magnificent adventure; the inevitable return in three weeks to a land she loved but whose people, revealing the ugliness of separation and self-interest by voting Brexit, she no longer knew. Then there was her frank attraction to Arun's feline self-assurance, like a cat he was incapable of a graceless movement, not to mention an animal magnetism she knew him unconscious of, how everything he said came pristine and curiously complete in the simplicity of his expression of it, the many sound qualities she was discovering in him that made everything right. The muddle of feelings came out as: *Oh Arun, I don't want to leave India, ever, I have been away too long, I am home here, and I missed you even while I was with the best people I have ever met ... I, I don't really know how to bring up the words of my feelings ... such old old memories, Far Memories ...*

He handed her his sparkling white handkerchief, a Scarlett O'Hara Moment – she was right about his sound qualities. They were a constant surprise and delight, no wonder his wife loved him. He waited while she dabbed her eyes, her cheeks, sighed and looked at his face with all the love the world could hold.

You are Hindustani, he had said, *your heart is Hindustani*, but: *That doesn't help me stay,* she smiled, ruefully, and, knowing how tears crackled old faces, turned away.

Come, come, come, he said, gently indicating they would walk back to the market, he would buy her momos: *The best in Nainital, but we must hurry a little, the kitchen will close ... timing, you need proper food.*

After momos, two shared plates, coffee, almost ossam, and selfies by the lake with a group of dizzy young Delhi belles who insisted – Arun stood with them all, film-star erect, a Leo with his pride of female admirers. Anna felt her heart just might break but it was too soft with love and only wobbled like a shaken blancmange. They stood together then, in their own mandorla of pleasure as darkness fell around them. Arun looked at her, up and down, quizzically, a pronouncement (one of those pristine pronouncements) on his tongue: *Well, you can't go to the Gala dinner dressed like that!*

Anna blinked. She was known for cutting a sartorial dash, but not, apparently, good enough for Arun whose preferences ran to *bling de rigueur*.

You looked gorgeous yesterday evening, he said, and: *Oh why hadn't he said so then?* Anna would have worn her grey silk again, with a different silk scarf, different earrings. She was touched that he noticed these things. She was, after all, *old*. His personal appraisal of her mode as not quite suitable for this gala evening, despite the addition of the handsome hand spun Kumaon shawl he had helped her choose the previous afternoon from Ram Lal Brothers, founded 1901, and which women had commented on all through the day, was not, *um*, up to standard. At least, not his standard for gala dinners. *Arun*, grinned Anna, charmed, *it's too late to go back to change and anyway, I've packed. You want to leave at eight in the morning.*

Her hero no longer needed to say, *come*; she would have followed him anywhere then. He led her to a perfume store and left her to top up from an amalgam of delectable testers of natural oils and unfamiliar fragrances while he disappeared to return five minutes later with a splendid pair of large oval amber earrings to match the amber pattern of her shawl's weave. They were perfect. *He unbends me*, she glowed as her heart overflowed.

The evening ghazals of exquisite depths wrenched her heart like hail bursts and roses in the soft garden lighting. Magical. Arun would not accompany her of course, the *of course* underpinned Indian society, but once she felt replete with food and compliments, had said her farewells to Siddhartha and Jaideep, shared last stories with Suman and Kavita, and especially to Janhavi who introduced her: *Meet Nigelle, our own Judi Dench!* she called him to drive her home to Abbotsford long before midnight when coaches turned to pumpkins and glass slippers became daylight's cinders.

the aureole moment by the lake

Leo with his pride

7

Coming Home

8th October. The Day Anna would remember, *Forever*. The house was silent, she could have heard an owl breathe as she padded down the steps and over to the main house of Prasada Bhavan for breakfast in the chandeliered dining room for the last time. The gardens were being cleared of their chairs, the rows of rainbow prayer flags already rolled away. Kanta had left for Delhi at dawn; Janhavi was overseeing the dissembling. A quick breakfast, a sandwich made for the journey, bags carried down by the houseboys from her loved and golden room, a handsome tip for all left in an envelope on the bureau and Arun arrived, a hasty: *Bon voyage, see you next year!* a final photo, and they were away.

The next part of the journey was pure pilgrimage, to picturesque Ranikhet first and the ancient Jhula Devi Temple, a temple of bells the Goddess was said to ring to announce her presence as protector of animals and healer of ills. Outside the town and high on its own hill Anna knew from her research that close by was a series of rock art paintings unknown to Arun but that she very much wanted to see. As they came from the Temple she saw to their right a set of steps cemented into the hill, ascending through light forest. Putting on their sandals the pair set off, up and up and round a bend to see: *Oh my!* two great boulders splendidly painted with scenes of Rama and Sita, Lakshman and Hanuman – the Holy Book of holy books for Hindus, telling of the rescue of Sita from the clutches of the demon rakshasha, Ravanna! Anna's story, fully present as Sita Devi!

Pulled by invisible threads she walked to the marble steps at the base of the rocks to sink down beneath and between the paintings, underneath the name of Rama writ large in Sanskrit, the lettering a deep terracotta ochre, above her. Anna radiated an inner light, Arun read it, photographed her. Her unfailing intuition for such things had once again dictated her day's ensemble from her tiny carry-on Tardis. Her charioteer noted it. The chiffon duster-coat, charcoal with creamy pink faded peonies splashed over, flawlessly colour coordinated with the patchy charcoal and creamy pink rock face. Arun shook his head, grinning, even her terracotta scarf picked up the Name, Rama, above her head, Rama the divine consort. The Devi Herself demanded the dresscode to meet the Divine. It was a woman thing. But more than that, She had called them to a sacred place not even he had known.

Were they tears Anna detected in the eye of her own godling consort? Finding these paintings was reducing them both to One of Those Moments. Tension held them in a thrall of enchantment, Arun's wonder that Anna should know of this place when, for all his visits to Jhula Devi, he had not known its existence. She was magic; this woman in his care was truly magic. She was speaking to him: *Oh Arun, thank you, thank you so much for this moment, it is such an important moment for me ...* she could not assume it was so for him, but: *No!* his voice burred with emotion: *it is very special for the two of us, both of us. You are most truly Sita Devi ji.*

He reached for her hand. She gave it. At his touch warmth on warmth pulsed through her, erasing their separate selves, a story dissolving and forming, dissolving and re-forming. He still held her hand as together they ran up the steps behind the rock art to the Mandir itself at the top the hill. They stood close as he lifted a prayer book from the stand and chanted. His voice filled her, shook the trembling edges of her skin. They walked clockwise around the inner sanctum. Outside, Arun asked a small group of young boys studying the holy books if one of them would take a photo of he and Anna together down by the rock paintings. All the children followed, and the slightly blurred photo would become Anna's most treasured as she sat on the step and Arun sat close, leaning into her. The dreamlike moment embraced them:

> *Remind me again—together we*
> *trace our strange journey, find*
> *each other ....*
> *Some time we'll cross where life*
> *ends. I'll touch you....*
> *Stars will move a different way.*
> *We'll both end. We'll both begin.*
>
> *Remind Me Again -*
> *William Stafford*

You are truly Sita Devi ji, Arun repeated, and had anyone asked Anna at that moment she would tell you, she was.

They found a lonely and charming guesthouse which promised lunch at midday, and a rather special ice-cream which, though Arun swore he wouldn't eat, at Anna's embellishment with fresh cold coffee and Hershey's chocolate sauce, he loved. It was their driver's birthday, Anna invited him to join them for a birthday lunch, the ice-cream was the birthday cake. Reluctantly, for Anna, they left by two-thirty at Arun's insistence to continue onward to Kausani. She had discovered a lounge of refined enchantment up the wooden stairs above the dining room, would have moved in right then if she could have claimed it her own with its corner devoted to a small porcelain Ganesh seated behind a tiny vase of fresh flowers.

The road was fraught with those *I-really-ought-to-panic-but-what's-the-point* moments. Driving on the chasm side of the narrow mountain road vast lorries came at them, overtaking each other, two abreast on perilous escarpments. Anna held Mrs Tweedie's title, *Chasm of Fire*, as a mantra in her thoughts and a talisman, trusting they would arrive unharmed in body and soul. A familiar peal and the cover of the very book itself appeared courtesy of WhatsApp from Jennifer, reading it in London as Anna was travelling to the retreat in Kausani. Another peal and Heidi in New South Wales sent a photo of a memento Anna had long forgotten – an envelope containing lavender picked for her from Mrs Tweedie's garden twenty years before. The timing of each synchronicity Goddess surprised; God sent!

By late afternoon they reached the lower village of Kausani, having stopped occasionally on the way to take photographs of the closing distances hand-painted on milestones along the road. Anna told herself the photos were to share with the meditation group in Glastonbury which had grown out of Mrs Tweedie's long ago London group but which, over the years, had turned to revere Llewellyn. A vivid dream of Tweedie eighteen months before had instructed Anna to return to meditate with the group after how many? twelve? twenty? years. She was quietly dismayed that Tweedie was rarely mentioned, and never named in the prayers-for-their-teachers before meditation. Only Llewellyn and his son were named; Anna silently sounded Mrs Tweedie, Guruji – Bhai Sahib. Beloved, ever-present, Mrs Tweedie, whom so many of the current group had never known.

After the crisis occasioning that dream Anna began to read again the magnificent Tweedie of later years in lectures brought together in a privately translated and photocopied edition of nine A4 booklets. She was privileged to own a set; Jackie had sent them volume by volume to Australia when Anna had returned there. *This* was the Tweedie she had known over four years, uncompromising, powerful, wise beyond the world, drawing attention to the riches of India, of Jung, of Teilhard de Chardin, of Catholicism in a way far-surpassing Anna's memory. Now, with the interim thirty years between, all was illumined in Anna's mind by words which sprung *living,* from and in Tweedie's voice, a voice which came from the Void to affirm her own journey. Tweedie had burned her presence by pokerwork into Anna's heart. India and its sacred teachings was the subtext of her life, Catholicism, a regular-ish outer form of attending to the Divine deepened by Father Bede Griffith's presence in her life. Respect in those great traditions meant naming the preceding guru's and teachers and the whole litany of saints with reverence. The practice took *seconds,* and shifted the mind to a different perceptibility, reminded the hearer of the continuity beyond corporeal death: *They are always with us, they hold us as we remember them. Saying their name opens a portal to a link in Consciousness; our attention is held in and to the moment. They are our ambassadors to the Absolute.*

Anna questioned herself being at the group. She knew herself an anomaly amongst the adoring, but, she sighed to Self, she had chosen the Cat-Who-Walked-By-Herself rôle of this life before she was born; permission was not hers to join or leave until 'told' so. She continued attending the Friday meditations, in silent gratitude to the beautiful couple whose home had been open to all, on all the Friday evenings since the very earliest days. Even as she took the photos of her pilgrimage to Kausani, to Mrs Tweedie's being there, Anna's intuition told her only Marion would be interested.

Kausani was the lodestar of her road trip. When The Mobile Phone magic'd to life and WhatsApp opened its wizardry, and Arun re-appeared to say *hi,* with roses, and he would take her wherever, which led to The Journey, Anna grasped at straws and found them solid. *Could we go to Kausani?* she had asked. *Of course,* he said, never having heard of it. And here it was, a tiny hamlet in a high valley, at the top of which sat an old, old ashram. Gandhi had lived there for some weeks and Mrs Tweedie, after Bhai Sahib died, went there for three months. Her letters at the end of *Chasm of Fire* begin in October 1966. It was now October 2018.

The tiny village was home to another extraordinary woman too, Sarla Behn, born in London in 1901, one of the two western disciples Gandhi named his 'daughters'. Her Swiss father and her English mother gave the young Catherine Mary Heilman a dysfunctional upbringing propelling her escape to India and to the Mahatma, who had advised Kausani when the gruelling plains of Gujarat overwhelmed her. Mrs Tweedie had stayed with Sarla B., had written of her: *A most impressive woman.* Arun was soon to be impressed too, Sarla Behn had been frequently imprisoned by the British for her Gandhian sympathies to Indian Independence, activities which ran counter to the Raj.

The sign soaring above the road read *Kausani .3 kms,* Anna asked the driver to stop. The ashram, she knew from Tweedie's letters, was higher up the mountain than the village, but it couldn't be that much further on, could it? She would walk, as befitted her arrival as a pilgrim. Arun would not leave her, though the way was unknown, the village was unknown and he'd certainly never heard of Mrs Tweedie, but his magical charge needed his protection and he would walk, *walk*, even the thought tested him, beside her up the towering hill where she sensed the ashram would be. He sent the driver on. The road wound round and round and higher and higher. Arun bought a bottle of water at a roadside stall, they needed it. On and on, as always when walking an unknown route it seemed endless; the following day as they walked downhill they would know it as barely a kilometre. At the top of the hill they came to steps, indicating a steep rise to the ashram proper. It led directly to an ancient bungalow in which was housed the modest reception with its inevitable library of ledgers.

Anna was given forms to fill in triplicate confirming her newly-discovered-since-Brexit Swiss identification and English domicile, the swami overseeing reception then pointed to dormitories running at right angles to the peaks. They had been built since Mrs Tweedie lived there, with no view of the mountains of which she had so inspiringly written in her letters home.

Seeing Anna's face, Arun, in voluble Hindi, words rolling like waterfalls over rock in their descent, explained to the dour-faced men Anna's reason for coming, for making this long pilgrimage, her need to be among the same marigolds in the same garden as *her* teacher, sit on the same steps as *her* teacher. Anna sat begum-like on the threadbare sofa and smiled graciously throughout their exchange. Reluctantly the chowkidar indicated they follow him backside to a path leading to an old and shabby cottage containing three rooms, with deep green doors and low steps to sit on, set in gardens of marigolds and autumn flowering roses and – *Nanda Devi!*

Anna knew by the tingling in the marrow of her bones that *this* was the very room. The chowkidar wrestled long unused padlocks to open shuttered doors to large rooms, which by the look of things, still held the imprint of Tweedie and 1966. The grime didn't bother Anna, the mustiness didn't perturb her, the damp – well, what's damp? Not even the too short, by at least four inches, hose attached to the sink drain, assuring water flowed out all over the floor, would deter her, nor the suspicion that the bedding had seen other bodies without interim soap and water. Anna was entranced. They were given the keys for each of their padlocks. She was *here*. Mrs Tweedie had sat *here*. Nanda Devi was *there*.

Her Hero of the Moment had done it again. Anna stood by the garden wall separating the cottages from the courtyard to gaze at Trisul, its three peaks sharply visible in the twilight. She watched as clouds rolled away to reveal the mountain Goddess Nanda Devi in all her magnificence, Queen amongst the heights. Anna stood on top of happy hours; it was a Holy Moment.

Supper called, the dining room was down long flights of steps, Arun held her hand firmly as they descended. Anna's varifocals play tricks in poor light. The kitchen, she photographed it discreetly – being Phoren and a Woman her shadow may well have despoiled the Entire Meal – was an astonishing affair of Fire and Floor. The food tasted marvellous. After supper the pair skipped up the darkened steps, Arun firmly holding her hand for her safety, back to their rooms. Anna wanted to upload the day's photos on to Candy, her very ancient pink plastic laptop, light for lack of battery, long since dead and no longer in production. *No-one,* reasoned Anna laughing at Arun's frank surprise at her State of the Ark computer, *would ever bother stealing this*. He agreed. But – oh, someone had banged a nail right through one of the holes of the only three pin socket in her room. And the casing was surrounded by a dangerous shade of smoke.

She knocked on Arun's door – could she use his socket? *Of course, come in.*

As she uploaded, downloaded, and whatever, which took a while as Candy was never a speedy click, Anna told

Arun again how special this day has been for her, his photos of her under the name of Ram, of them together: *They will always summon up the magic of my pilgrimage to Kausani in the years to come when I am no longer with you, no longer in India. They will remind me of a dream come true, and that you are responsible for it. I could never have done any of this alone.* Haltingly, shy now, she asked: *And do you have a special moment of our journey...?* tailing off as she spoke, suspending her breath as she waited for his answer in the beautiful pause.

Arun was sitting on his bed, lotus positioned. He looked away to the window, thinking. Anna smiled to herself, she knew the same gesture as her own, as if looking at yesterday and the day before to catch the memories before they dissolve into tomorrow or the next day. He was pensive, quiet, and stayed so for a long time before he turned to look at her, speaking thoughtfully, choosing carefully the words:

When I first saw you in March and you were shaking so much and those Indians were not looking after you – pause *– I felt something. When I take you to the airport I did not think I see you again. But now after all these days together I have not even thought of my family or my home. I am totally with you. Totally. All the moments are magic. You are magic. I feel it. You have such a pure heart. You are old, but you are so young. You think you are seventy-two, I think you are twenty-seven.* His look, intense, holding her eyes, measuring her heart, held back words he wouldn't think.

Silence. Anna was moved to silence. Words held in by the boundary of her heart would remain unspoken, suspended in the air along the songline of their gaze. The *bond;* he had spoken of it as they walked from the airport to his car when he met her from the plane. The magic between them would settle in its own time and into its own perspective. Anna knew their lives beyond this moment were and always would be separated by vast imponderables. She was blessed with his presence *now*, blessed by their meeting *then*. He wrote in her notebook:

Rama - राम अ Sita – सीता
Sita Devi सीता देवी

The photo back-up complete, Arun walked Anna back to her room, it was dark and no light lit the path, he was her chevalier. He asked to come in to take a good look at the smoke-stained socket – without it she would have no power to recharge her camera or her phone for the following day. He would be using his for his own phone. He assessed things, smiled, went to fetch a screwdriver and returned accompanied by their driver and two men from reception.

D'you need a torch? Shall I turn the overhead light off? asked an innocent Anna and all four heads nodded negative. The men were crowding the only door out of the room, too occupied to note her apprehension. *We need the light to see,* said Arun beaming confidence in her direction. She walked round the bed to the furthest corner of the room, took a photo of Arun and his Sparkies as he dug his screwdriver into the live socket to remove the cover to get at the firmly embedded two inch nail.

Anna took a quick photo on her phone, began tapping a message to accompany her WhatsApp to friends in Australia and England; *If you don't hear from me again this is why! Arun is removing a two inch nail from the only socket in my* room. *The power is still on. He needs to see what he's doing. The other men are watching.*

Within seconds her phone rang a crescendo of peals in response: *Are you alright?! Good heavens are the men okay? Are you still alive? Are they still alive? Can't you stop them? Good grief, is Arun* smiling???? Deborah, Jeanne, Heidi, Steph – each responded in concerned astonishment.

With all uncovered live wires sizzling, the cover was off. Arun hammered out the nail stuck in the hole, a loud clatter as metal hit tile. With caution and sensitivity, Anna noted, he tentatively re-folded the wiring in its place, screwed everything back and even more tentatively tested the switch. Voila! Her Hero grew Apace! The audience of men left; she had light and could charge her camera safely.

Laughing in relief, she replied to the phone messages: *I'm breathing! My Man of the Moment has done it again! Rescued Damsel in Distress, brought Light into my Life and Gone to his Room a Happy Hero!*

8

Bathos!

Anna went to close the heavy vertical outer double doors to her room as the men left. There was a knack, she thought she remembered it from French doors, in France. The edge of one door was concave, the other convex. Closed to almost flat, the two edges would slot into each other and a hefty push locked them tightly shut. She pushed mightily against the right panel – and almost passed out. With a force so swift and too spring-loaded to move her left hand positioned on the door panels at the two edges of overlaps, it sprang shut, *fast*. Closed, despite her hand. Spring-loaded, the doors did not open. In anguish Anna pulled with her right hand the handle of the now firmly locked-in-place wooden doors and gasped with pain as blood spurted from her released palm. A Whole Drama of blood and pain. In agony she ran up the path sobbing: *Arun! Arun! My hand! My hand!* and banged on his door, helplessly. She heard his muffled call, he was in the bathroom: *I am coming, I am coming!* he assured her.

Opening the door he stood momentarily horrified at the red-black blood oozing from her palm with its rapidly discoloured bruising. *I crushed it in my door,* whimpered Anna, now soothed by his presence, comforted by his calm control of yet another situation.

Come, come, come, he said, seating her on his bed while he rummaged in his duffel bag for another snow white handkerchief and a first aid kit. Anna's accolades for Arun were increasing exponentially, *hourly,* as he folded the handkerchief precisely in a wide diagonal to wrap around her hand, sopping up blood as he looked for ointment. Not finding any he rushed off to the office, returning with something that made Anna cringe. The cap of the something was so firmly fixed the contents had been squeezed through a crack in the filthy body of the very well-used tube. *Poor loved hand,* thought Anna, *you'll survive through Arun's compassion but probably rot with gangrene before the week's out.* She sighed and hoped love would win.

It was so, so, so painful. After Arun's doctoring Anna returned to her room under his solicitous escort. She wished him goodnight, blew a teary kiss as he closed the doors from the outside. She pulled out her photo of Tweedie, rested her hand on it a while, prayed and actually managing to smile at the day with its holiness and humour, pleasure and pain, experienced in sequential perfection.

And so to bed, and ... *My hand is most comfortable, as they say,* she WhatsApp'd the latter-day drama with a photograph of her hand, Bound and Betadine'd by himself and now resting on Mrs Tweedie's photo to aid its healing process, to her friends. Tomorrow was Another Day in Kausani.

The Hand was an Issue. Constricted by lubricants and bandaging, lovingly changed twice daily by Arun, within two days the wound was becoming swollen, a septic yellow appearing at its edge. Anna thought it needed air.

She had no need to lift or carry anything more than a pocket camera in her right hand as Arun looked after Everything. She peeled away – many wincing ouches – the bandage, the tape, the cotton wool – real ouch there, fibres had stuck fast in the blood'n'Betadine – and, notwithstanding she couldn't uncurl her hand as the wound would re-open, left it uncovered to breathe. At night she covered it with a bandaid for courtesy and safety, bloodletting on bedsheets that she wasn't washing seemed a poor way to repay a host, whether hotel or ashram. It took another ten days to heal sufficiently to more or less open her hand, and a further week before she could use it without pain, but fresh air worked. Still, it would be a long fortnight before she could uncurl her palm.

Morning weather, edged with chill. Anna rose long before sunrise to watch Nature's gift of rose on snow, warming herself with chai from the chai vendor, whose daily pre-dawn appearance in the carpark in the shadow of one of the glories of world was a benediction. Arun, muffled in scarves, joined her as she sat watching a world wake in beauty, silently warming her hands on her *teen* chai, please' tin mug bought from ProperJob days before she left.

After breakfast in the ashram-bleak dining hall Arun came into her room for the large blue bucket, he would return with it filled with almost boiling water for her bath, a one-handed operation which, with her hand hampered, rather left comfort out of its pleasure. Washed, sort of, but dry and warm again Anna came out into the sunshine to see two chairs placed in the garden, facing the snows. Arun's anticipation of her every thought set

Anna's mind to wonder. Her whole life – her *whole* life – she had dreamed and day-dreamed of a man who would care for her in this intimate way. *It's worth being an Indian wife to have such attention,* she sighed to Self, knowing that in her whole life – her *whole* life – she would only have three weeks of it.

The pair sat together in the chairs, watching the snow peaks, wisps of smoke from early morning kitchen fires curling up from the forest below them, the great deodars, trees of the Gods, *deodar,* Nature's magisterium, silent. Anna held photocopied pages of Tweedie's letters in her good hand: *May I read these to you?* she asked tentatively, she didn't wish to assume, or to impose on his attention to the mountain. *Please, please,* he replied, and Anna began:

October 1966
Dearest
This letter comes to you from a solitary retreat in the Himalayan hills. I am writing seated on my doorstep, facing the snows. They are clear this morning. And last evening too; the whole range was coral pink, the glow after the setting sun dying gently away on the glaciers.
And so near they seem...
It is a glorious morning ... Our ashram garden looks like a valley of flowers just now ... All around are high hills, the Kumaon Hills, covered with pine forests ... It is perfectly still, then suddenly, as if obeying the signal of an unseen conductor, the birds begin to sing ...

At that instant a bird alighted on the branch above them, pouring out its song to the Beloved. Arun looked up: *She hears you,* he said. Anna paused, looking up at the singing tree ... continued:

> ... *the Sound is deep, endless, eternal. The yogis in Rishikesh say it is the Nada, the Breath of Brahma, who can never sleep ...*

Anna's voice caught, tears shone on her face in the morning light, more words would not come. Arun reached over, a grace-filled gesture, to take the pages from her hands. He continued to read to her, while the bird sang above them: ... *can never rest, otherwise creation will disappear into Nothingness.*

Arun continued to the end of the paragraph and paused. *I have not heard the Sound of God,* he said in his thoughtful way, and Anna shared with him her experience at Arunacala forty years before, describing it as the song of the bird above them reached a crescendo of purity, its song falling in vision before them both. Sound become vision – isn't that surely the Cosmic Reality of all faith, all *faiths,* all traditions? In the Beginning the Word; In the Beginning the Cry; In the Beginning: *Kun!* In the Beginning: Vāc the wife of Vision, the Mother of Emotions, the friend of Musicians. Cosmic Sound. In such icy solitudes legends live in the Himalayas.

Dissolving, mutually, the tension that held them Arun set down the papers as both sat back to listen to the bird just above their heads. Its song complete, it flew off. The pair rose, Arun replaced the chairs from wherever he had found them, and together they walked the kilometre down to the village where Anna bought a new tube of Betadine, bandages and bandaids, gauze and papaya. Papaya was her preferred fruit, enjoyed with *dahi* she could survive any crisis. She comforted herself with the thought of its death-defying enzymes and wished she had included a tube of Ozzie pawpaw ointment in her tiny carryon for applying to The Wound as it closed over.

After lunch the plan was to walk over to the opposite hill to the ashram and *samadhi* of Sarla Behn. Each of them had respects to pay. Beforehand they would explore the museum dedicated to the life of this formidable woman for whom Arun was as much taken as Anna. He wrote in the museum visitor book of his appreciation of this Englishwoman who locally played so much a part of Indian Independence, and suffered for it.

The days in Kausani were full. Anna was Full. It was time to leave. Anna had achieved her dream; a thought hardly possible six months ago. Her gratitude to Arun overrode her commonsense and would continue to do so long after she returned home. She trusted him with her life.

The following morning they were sitting on the low wall in the garden, companionably close, quiet, when Arun's phone rang. *Excuse me,* he said, and rose to walk, then pace, with mounting force in his voice, up and down the path in front of the cottage, up and down. Anna understood some things, his Hindi included sprinklings of English. She thought he was speaking to his daughter, his rising voice addressing money, demands, school fees. He was annoyed. She should be at school. The money he was making, this was a *working* trip, would pay for her school fees. Then something more about the school, not having received ... but she must attend, he would phone ... Anna caught the gist. Her heart dropped like stone. Would this one phone call colour grey the rose of her journey?

It could, and it did. Arun's remained distant, back in his own world. Anna didn't belong there. At that moment, reminded of the impossible number of kilometres calculated at the beginning of their journey, she wondered if she was simply a walking wallet for his daughter's, his family's, fiscal affairs. She remained apart, closing in to herself, treasuring the moments of the past days which would never be gainsaid.

The next destination was a full day's drive to Rudraprayag, a wild journey along treacherous roads dug out of sheer mountainsides, granite and mud, now washed out by recent monsoon rain. Landslips of whole hillsides, rock falls and chasms where once were roads, Anna should have panicked, but reckoned it pointless. They would survive – or they would not. The hundred? hundred and fifty? kilometres walled by the wonder of the majestic snow peaks of the Himalayas as they drove bend after bend kept her life-down-here-on-the-road in its true perspective.

Crazy driving by oncoming lorries tottering on the mountain edges necessitated unusual caution for most, but not all, drivers. Road blocks proliferated, one stretch of eight kilometres took two hours. And food, Anna needed food of course. Carrying bananas and papaya had its limitations as two hours became eight hours, stretched the calculated four and half hour journey to ten hours, unrelieved by restaurants along the road slips and landslides.

It began to rain, the first hefty drops turned the sky leaden, swiftly became liquid carnage. Mud roads turned to porridge. Anna needed to eat. Miraculously, enterprising men had erected kerosene-tin pop-up pit-stops in lay-bys along the way. In one Arun rummaged around to produce – Amul Lassi! A fabulous yoghurt drink which would support Anna through some of those hours. Lovingly hand-patted momos were prepared at one wayside stop and cold samosas sat heaped on a counter at another. Yet another, even more doubtful than those previous, offered dubious food which Anna ate gratefully, shutting her eyes to its sourcing. The washing up was mountain water piped by the road side across the porridge. Anna ate Everything Offered, certain that love was stronger than the hygienic truth. She survived it all, fit and well.

At the Literary Festival Stephen Alter had suggested a route she should best take from Kausani and the best place to stay in Rudraprayag, and, mildly contradicting Arun's desire to go a little further to Chopta, a small town he knew, and encouraged by his apparent defection, Anna quietly insisted they follow Stephen's advice. She was, when all said and done, paying. A thought she kept to herself, but which would henceforth reactivate a certain independence of spirit, despite her sorrowful struggle with heart.

The three were exhausted as they coasted in to the bustling town, the driver had proved his skills on that road. Arun asked a local for directions to the Government

Rest House. Pointing up a one-way parallel hill Arun indicated he would walk and the driver would continue on to turn back onto the higher road. Anna walked with Arun, her legs thankful after such a ride to find strength returning in action. A longish walk, but at the upper end of the hill they found the Resthouse and ... were enchanted!

Built on the upper slopes of Rudraprayag town all the rooms of the GTRH had balconies suspended over deliriously scenic views of the *prayag*. It was mesmerising. The confluence of the two great Rivers, Alaknanda and Mandakini, merged exactly here to become the one mighty Mother Ganga. Rudraprayag was one of the Five most holy Confluences: *Panch Prayag*. Flowing in from their mountain sources, the jade green Mandakini met in turbulent spume the steel grey force of the Alaknanda. From their confluence right under their balconies began the holy Ganga, in a tumble of white foam. Over breakfast, lunch and dinner in the dining room, which windows suspended over the confluence, Anna and Arun sat silently taking in the spectacle that powered Indian prayer and thought. Anna was spellbound; Arun too. He had not known of this Resthouse, had never seen this aspect of the Prayag: *It is perfect,* they agreed.

A night and the remains of the day, with an hour or two in the morning, was all they had at this special place, barely long enough to dream. In the early morning mist Anna, dressed in her long misty grey silk, stepped out on to her balcony, a Juliet Devi moment. Suspended over the gurgling waters below, the sight inspired an unsuspecting photo by her Romeo on the adjacent balcony. Arun shook his head, she had done it again, unconsciously colour coordinated – the early morning mist around her, silver grey; the river, below, turbulent, silver grey. The photo, taken on his phone in too little light, proved too out of focus to enlarge. Anna thought it an apt metaphor for her, back on the periphery of his world: a world which had momentarily held them in its magic beyond the mist.

Reluctantly, Anna made ready to leave. Arun said he would have liked another day, and would return. Anna's heart dipped, she would likely never pass this way again in this life. They left the magic to continue another whole day's drive, another torturous eleven hours, via Tehri Dam to Landour. More landslips, more road blocks, more daft driving, the bride in the dollied-up car coming at them from around a huge lorry going too slow for their driver's liking, almost didn't get them to the church on time.

    Shortly out of Rudraprayag Arun told the driver to stop. *Come, come,* he waved to Anna, as he crossed over to a small gopuram gateway and then waited as Anna looked down in wonder at the sight far below. *This is Dhari Devi Mandir,* Arun told her, *she is rising from the river itself. She was lost in the Kedarnath flooding, and is all new since then.* Architectural wonder it was indeed, the temple rising from the river and only accessible via a long, long footbridge angled from the bank to the temple door.

    Arun led the way down, down, down, past chillum smoking swami's with feet bound in rags, and women in wondrous shawls; past stalls filled with malas and marigold garlands and holy trinkets, where Anna bought a tiny plastic pendant of Durga Ma on her Tiger.

The temple to the Goddess was to be demolished on June 16, 2013, to give way to the construction of the 330 MW Alaknanda Hydro Electric Dam. Hours after the idol was moved flash flooding from freak thunderstorms and exploding glaciers upstream caused the country's worst natural disaster. Devastating floods and landslides washed away the entire shrine town as well as flattening Kedarnath further up the Mandakini. Uttarakhand had to face the ire of the Goddess – and the 330 MW hydel project was left in ruins. Millions, *millions* of £'s, $'s, Rs gone in the raging floodwaters. And lives, hundreds of thousands, maybe a million, human, animal, everything living, swept away. A similar attempt was done in the year 1882 by a local king which resulted in a landslide that had flattened Kedarnath. One doesn't mess with the Goddess.

Inside the Temple no photography was permitted. Anna sat very close to the Image of the Goddess, a stone face black as pitch, there were two, each garlanded with marigolds and gold tinsel edged red gauze, fitting for Dhari Devi, herself a Kali manifestation. Anna found herself unaccountably moved to tears; just as some of her beloved Black Virgins had enveloped her in their arcane power and mystery throughout France during the 1990's, so did Dhari Devi now. *I have seen You, Great Mother,* she whispered, *with the eyes of the heart.* It had been a long journey to The*a*logy.

    Arun sat to her right, slightly behind. Anna's camera was on her lap. It didn't occur to her to use it; she was mostly respectful of requests not to photograph, and besides, behind her were a dozen eagle-eyed priests. But...

Without warning the four priests serving the Goddess, behind the grill that separated the Holy of holies from Anna, rose in a body as if by Divine Command and disappeared behind a door on the left. *Quick!* the Devi commanded Anna, *do it!* And without even pausing to wipe the blurring tears from her eyes – she did as she was bid! Arun saw her low sleight of hand and was appalled. But also elated – he would have his own image now for his personal puja. Anna assured him the priests had moved (most unusual, he could not gainsay the fact) at Her Command and She had whispered, lickety-spit: *C'mon Sita Devi, there's no Second Chance.* A soupçon of smugness made Anna smile. She rang the rows of Temple bells in gratitude as she left with her tangible darshan of Dhari Devi.

Ooops, head on ... really-ought-to-panic-but-what's-the-point ...

9

Tibet, Tibet

Heading north from Rudraprayag the road was reasonable as far as Tehri and the dam that flooded the village, beggared the hills of their forests, the land of its life – leaving a seismic belt of shale simmering with stress and fracture from blasting, potent for floodwaters from snowmelt and glaciers upriver. Arun stopped the car at a point and peered, from under a pointless three foot square board with faded lettering warning No Photography, at the dead scene spreading as far as the eye could see. Little wonder the Goddess was angry.

The two hundred and sixty kilometre journey seemed nightmarishly never-ending as they wound round and round and up and down a landscape in turns desolate and dazzling, heights and valleys, riverbeds and peaks. The road was fraught with more of those: *I-really-ought-to-panic-but-what's-the-point* moments. More rain, heavy with hail, slowed them to a crawl. A couple of chai stops and a small café on the outskirts of New Tehri put to rest Anna's food concern for a couple of hours but it would be eleven hours before they, she ragged and faint, arrived in Landour.

Landour and Mussoorie; the mountains were ever-present, whether cloud-shrouded or not, they were there and the air felt pure and clear, the forests rose with them as they rounded bends to climb higher; bird calls echoed, rising away into the forests, wave on wave into the distance. And those peaks! A strange wild yearning swept down from those peaks, snow white and patched with mysterious shadows compelling Anna to look, look, as her heart – *ached*. *Ached*, with that yearning for Something ever beyond, stirred by the haunting call of geese on a chill autumn night when in long arrowed skeins they left the Somerset Levels for warmer wintering ground, flying high, leaving her behind them. She took the loss personally, she so often wanted to go Home. The majestic Himalayas had always been home to sages and pilgrims, the Rishis wrote their verses inspired by them, the Gods lived there, the Goddesses. She could never tire of the Himalayas.

Anna had booked two 'deluxe' rooms for two nights at Domas Inn before she left Glastonbury. Their online sales pitch enchanted her; refugees from Tibet had created a haven for themselves in other mountains after the horrors they had left behind them a couple of generations ago. The owners had, over the years, commissioned an artist of exquisite talent to cover every square inch without and within the modest building with that glorious vibrancy particularly Tibetan. The photos on their website enchanted her. What a special treat, she thought, for a travel-worn pair to experience *deluxe* and mountain views on their nearly last days.

    Arun chided her After the Event: *The story,* he said, *it is an art, you believe it and you buy it. It is all lies.*

Not all Tibetans glow with the integrity of HH the Darling Lama. Wily, not wise, could be said of many governed by more mercantile interests. Westerners, sympathetic to their exile, weave rosy fantasies – Patrick French was right about that.

Anna did buy it. That's what good marketing is all about – stories, telling you just what you want to hear. Mountain views? A handspan beyond her upstairs window, the only window, was a mountainous building site that had stood unfinished for *years* prior to her booking. She asked particularly for mountain views, or at the very least, *hill* views over the valley, it was her treat for Arun and herself; two days of pure tourism, pleasure without purpose.

Arun and the driver, with Anna, older and with physical limitations, wretched with tiredness, all needed food. She had the wit to ask to see the rooms. Hers was inspirationally seedy. Seeing the vast building site, a hand span across from her window blocking Everything right, left and centre, and the indescribable bathroom, knocked what grasp of focus she retained after the long drive. Arun's room was purpose built, new with pleasant facilities, no view whatever, but it was outside and up the street away from Anna who would be left in the Inn completely alone. At night the owners went home.

The owners of Domas spoke excellent English. *We will leave,* said Anna, *I am so sorry, but your marketing said beautiful views.* Arun agreed. But the owners saw her state and brought her chicken soup, and it was very good and she was so exhausted and where else would they go, so late?

Deadening the warning bells clanging in her tired head Anna caved in against her infinitely wiser instincts, Arun's room was pleasant and he was tired too, and so they stayed. They would move tomorrow.

Her bag was carried up the half spiral staircase; another hefty quilted bedcover was brought in for warmth. She went gingerly into the bathroom. The towel was dirty, not just used, dirty, with an unidentifiable brown lump stuck to it. She asked weakly for a clean towel, one came, minus the brown affectation but scarcely Savoy standard cleanliness. The tap feeding water to the loo was off at the wall. She turned it on. Flushed.

Water and effluent rose up as she watched, helpless to stop its rise and rise flooding the floor. She closed off the tap in numb-with-tiredness disgust, cleaned her feet using the single tap, cold water, at the squalid and filth-mired sink.

Too tired to do anything but fall into bed, tomorrow she would leave. A knocking, through the wall of the apartment next door, semi-prevented sleep for an endless time.

Anna woke too early, sugars rock bottom, went to find someone to bring food. The front door was unlocked, no-one had stayed in the building, Anna was sugar-panicking, she found the kettle in her room broken, not plugged into any of the five available sockets would it work. Shaking, she went back downstairs to hunt for another kettle, found one in an airless windowless room obviously not classified as deluxe, took it back upstairs, by now almost sobbing with stress and sugar-drop. She had a banana and her Lapsang Souchong – could she last two, three hours? Sugar panic overwhelmed her when she realized how alone, isolated and in such a horrid horrid place, despite its stories and gorgeous decoration, she was. She couldn't flush the mephitic toilet and the floor was too fouled to shower in with bare feet.

Tashi had given her his phone number – in case. This was a case. Anna phoned and phoned. There was no response, nothing. Now trembling uncontrollably Anna knew her body was slipping into serious alert; yesterday's erratic food intake and only a liquid soup late evening had done little. She ran barefoot and trembling up the street to Arun who took one look at her shaking state and ordered her in to sit on his bed for two minutes. He knew they should have left last night, those Tibetans were too *too* smart, they were reading Anna's weakness and they got it right. The chicken soup wasn't enough after that long day.

She calmed down now he was in command, they returned to the hotel. He was horrified to witness the unlocked door, her vulnerability at being in the place alone, the rank squalor of the bathroom and the irresponsibility of Tashi in not answering the phone. Arun thought on his feet, helped her pack, bellowed for someone, chowkidar, anyone, phoned Tashi a dozen times, received no answer. He took Anna's bag downstairs, his voice rising to a full-throated Leonine roar for: *Bhai! Bhai!*

A sleepy, surly creature rose from the gollum gloom in the depths of somewhere beyond Reception and Arun roared, *Omelette!* and some other choice words to which Gollum pointed to his watch, refusing. Arun's eyes and voice whipped the creature roundly, he lost no time in pointing out the peon's responsibility to Anna as a paying guest, and she needed food: *Right now*. Gollum sloped into the kitchen, Arun told Anna to sit down: *I am here*, he said softly. His presence stabilized her panic and, shaking, she could sit quietly until a thin, single egg, omelette of palest cream, appeared on a plate. Its colour told her it was not a happy egg, nor was the chicken that birthed it. Anna sent it a prayer of gratitude, gratitude high on her attitude list now Arun was in command.

He, all the while, had leant over the reception counter to take the Inn phone and repeatedly call every number on Tashi's card. Anna, sugars barely restored, walked shakily up the hill to find somewhere, anywhere, else.

Ivy Cottage. The chowkidar there was hardly forthcoming. There were no rooms. Well, maybe there were but the owner was away. Well, maybe he wasn't away but only sleeping. Well, yes there were two vacant rooms but they were nine thousand rupees each. Over ninety pounds, *each*, and Anna was paying for both. Overwrought and still sugarless, no amount of prayer could have willed much sustenance into that dead egg, Anna stood in the early morning sun and burst into tears. She told the chowkidar her horror story of his neighbour Domas. His whole demeanour changed.

One minute, he said, and disappeared to wake the owner. Ashish was erudite, intelligent and clocked the scene in one. He reduced the price by a half and walked Anna up the steps to show her two blissfully clean, large, parquet'd suites set in their own garden overlooking the hills, the mists, the forests of deodars shimmering in the morning stillness. Its perfection pierced her heart. She would return with her bag and her friend.

Arun took his rôle of chevalier seriously, and seriously expected a hero's recognition. Anna stood bemused, and a touch embarrassed, as he demanded a TV for his room. A *large* TV. He slept with one on. In that nano-second Anna's hero took on clay feet. She managed a laugh, at herself and her rosy fantasies. The chowkidar brought breakfast, setting up a table and chairs in the garden overlooking the forested valley. *On the house*, he said, smiling and then turned to lesser peons to install a huge TV in Arun's room. Arun took an unflattering photo of Anna as they sat in the sun overlooking the deodars in the valley below, too close for any comfort. She named it, *After the Tears*. A good souvenir, actually.

She had to return to Domas to Sort Things Out. She would pay for Arun's room and her chicken soup, his meal of the night before and her omelette. She would not pay for her squalid, filthy, viewless bothie with its mephitic washroom.

Tashi's mother was waiting for them, the chowkidar had called to tell her the two guests had decamped. There were huge eruptions to greet Arun and Anna as Tashi's mother's pretty face turned venomous. Really. She turned into two different people. Anna stood her ground. Arun stood *her* ground. He also took Tashi's mother upstairs to see the disgusting bathroom and toilet.

Great heavens! In their short absence Tashi's mother had cleaned up. And shouted at them that the kettle was working. Anna flew at her, anger ignited by injustice.

That does it! she fumed, *Arun witnessed the awfulness*. Tashi's mother jumped in: *I will call the police!*

Call them, said Arun in his calm way. She didn't, of course, call anyone, not even her hefty son. Arun's driver came in, entered the fray. He told the mother and the gollum he had known Anna for two weeks and never had she criticized even the bad places she had been or seen. If she said all was wrong here, then all *was* wrong.

Tashi's mother ranted. Another member of the staff appeared, an Indian as calm as Arun who was shielding verbal blows with: *She will not pay for her room. She will pay everything else but not her room.*

She booked two rooms for two nights, screamed Banshee.

She asked for views and a clean room, a deluxe *room, and a deluxe room must have a working bathroom,* said Arun firmly and quietly.

The threats extended to the ridiculous: they would do this, or that, or anything else, Banshee snapped sourly.

Do it, said Hero, *she is not paying for her room. Make up the bill.* Banshee's offsider, another staff arrival, did.

No, said Arun, pointing to their calculated attempt at extortion: *She is not paying for that room. And she is not paying for two nights because your story of views is a lie and we are leaving.*

Arun rose heroically to fight her pitch, Anna knew she could leave All in his hands. He waved at her to go outside.

Anna smiled obediently. Inwardly she mooned and swooned and went outside to sit on a Park Bench in the weak sunlight; she had a history of park benches, Park Benches in Romania, Park Benches on Patmos, all proved excellent safety valves for whatever local crises she encountered. *I forgive the TV,* she thought to Self, *if ever I dreamed of a Man Who Would Fight My Battles it is All Coming True – right in front of me.*

The battle raged for two and a half hours of that morning. Two and a half hours wasted. Two and a half hours Arun had stood quietly, heroically, refusing to yield one inch of his truth of the matter. Anna sat in the sun outside the Inn, she knew Arun's proud surety, grand and a little mysterious, would rightfully win her battle, Troubadour that he was. Repeatedly along their journey she had observed him as he had spoken with others, incredibly intelligent, confident, calm, masterly and curiously childlike: *A potent mix for perfection,* sighed moonstruck Anna.

The Indian who had gone in came out, sat on the bench with Anna. Hearing the fracas a worker from the offending building site came to sit on the bench with them; the driver came out to join them on the bench. There was no more room on the park bench and all set to in quiet tones to dissect the happenings inside the Inn. The driver told the men sitting beside him: *She is not like that, she is verry truthful woman, I know her for two weeks*. The man from inside the Inn nodded his head and said the owners were very bad and very *dürrty*. The walk-in from the building site grinned broadly when he heard Anna had booked a room with a view.

The anger ricocheting round the inner walls of the Inn flowed over Arun. He was enjoying his leonine moment quietly defending one of his pride. He hadn't the rose-spectacled view of Tibetans that he knew many Westerners in India held. He knew some of them were wily and ungiving when it came to commercial pursuits.

Anna was right, and he encouraged the *khro* to call for the police. He was amused that Tashi wasn't answering his calls, had sent dragon mother as the big gun. She wouldn't get past him, not now Anna had dug in her heels, he would defend her to the *enth*. Banshee didn't like uppity western women; she was used to fawning western hippies who fell for the Art of Pretending to be Tibetan Holy Buddhists.

Almost midday and Anna needed to eat properly. Arun and Banshee came outside, she, waving a piece of paper: *You don't need to pay anything,* the foolish woman said with loaded venom. Anna replied: *Yes I do,* despite the real temptation to say: *thank you very much,* and get up and leave. Arun looked on approvingly: this woman in his care had *class*. Banshee began again, Anna said: *Show me the correct bill,* checked it, knowing Arun would have already done so, pulled out her cash and handed it over.

In the security of loveliness up there in the garden of Ivy Cottage Arun smiled and reminded Anna again: *The story, it is an art, a fiction, and you believe it and you buy it. It is all lies.* Smiling now, she agreed, sighing to Self: *My dark godling, he can have his TV forever...*

10

Peter Rabbit

After the Horror Story of the night and the morning a lovely afternoon unfolded. Anna's fretfulness of the wasted hours subsided as the sun gained a little noonday warmth up there in the high hills and she relished breakfast and breakfast again at lunchtime. The first was a gift from the kitchen of Ivy Cottage, sympathetic to the state she was in. Now she was up for a Long Walk. Arun looked at her with his: *Don't even think about it,* expression firm on his face. He didn't do walks for the sake of walking. After his glass of hot milk to which, Anna observed, he added dollops of butter and half the sugar quota for Landour, all he wanted after That morning with the Tibetan *khro* was a good film on his TV and a rest. Anna left him to the heavenly views of the valleys in mist, the autumn dahlias in the garden, the peaceful ambience of Ivy Cottage – and his TV.

She walked off briskly up the hill a couple of kilometres or so, charmed by signs: *All those who wander are not lost*; and valleys streaked with shafts of sunlight, and when she reached it, the golden light falling on the warm timber pews through the long windows of St Paul's. Here was the famous row of four shops she knew so well from the whimsical writing of Ruskin Bond.

She watched a truly beautiful couple and their children arrive and sit at the café; the young American woman tall, slender, long blonde hair, the perfect paragon of her race; her Indian husband equally tall, elegant, graceful, the film-star perfection of a young Shashi Kapoor. Anna felt dumpy and momentarily discomforted in their presence, the morning had discombobulated her more than she realised. She walked on, up along the higher fork in the road, to the famous Bakery for a cappuccino which proved so weak she languished. It wasn't possible to eat any of the blow-torch sweet cakes they specialized in – English and Indians alike share a nursery palate for sugar – but she was greatly nourished by The Signs: *We do not have wifi. Talk to each other. Pretend it's 1895;* and: *No Smoking. If we see you smoking we will assume you are on fire and take appropriate action.*

She poured over the bookcase of Local Famous Authors, third generation western Indians such as Ruskin Bond, Stephen Alter, and Indian adoptee Scotsman Bill Aitken each of whom lived in this hill station of the forgotten Raj. She bought five books. And noted the wonderful shrine behind the counter as she paid – yes, it really did contain a ceramic Peter Rabbit! Anna fell in love with Landour; civilised, familiar, quaint, Clovelly cute. Walking, and her pleasure in fresh air, restored her.

Back at Ivy Cottage she found Arun up and seeking nourishment, and of the firm decision they would go down to Mussoorie Mall. Anna found its cacophony, its chaos, hell; she thought they were going to explore the Mall in Landour with its shops of curios and the antiquarian bookshop she had noticed on their way up, or was it down? so much had happened she was no longer sure. But she did know it appealed far more than the endless Mall of the lower town at this moment. She was more than grumbly, and needed *proper food*, as Arun constantly reminded her. He would, of course, win the toss.

As they entered the sheer pandemonium of the Mall Arun suddenly stopped the car on a congested bend and leapt out. People were shouting, pointing, he held up his arms to stop traffic, almost blasted off his feet by horns from a hundred vehicles and their deafening din. Swiftly he bent under the wheels of the car and came up with – a tiny bemused white and tan puppy. It had become entangled in the wheels of the chaos and, of all improbabilities, Arun had responded from within the car, got out, picked it up and carried it gently to place on the safer side of the street. It was unhurt. Anna melted: *He unbends me,* she sighed to Self, again.

Once she was out of the car, however, and in the mêlée proper, she couldn't control her failing temper – until Arun fed her hot boiled sweetcorn heavily buttered, with a sprinkling of real rock salt from the Himalayas, from a street stall. It was heaven. Once she was sugar-restored the chaos of The Mall became a manageable – *magic.* She could love it now, and she did! *This man,* she mooned yet again, *knows me!*

The bookshop sign promoting his latest book would be the closest Anna came to meeting Ruskin Bond, who, Stephen A. had assured her at the Lit Fest, would be open for a spontaneous visit. She had actually come to Landour wanting to meet him.

Before they had left for the Mall Arun and Anna walked up the steps that led to the home of the man himself, Anna tentatively knocked on the door at the top. Knowing how she felt about causal drop-ins herself she didn't feel completely comfortable with Stephen's assurances. Arun knocked, firmly. And each continued, respectively, with plenty of waiting time between knocks should the great man be in the kitchen, in the bathroom, or at rest. Eventually his so familiar Toad of Toad Hall face appeared through an internal door, they could see him through the glass, he growled myopically at them to go away. They went. Anna didn't blame him in the slightest; she was equally protective of her siestas and catnaps. She posed by the full sized vertical banner outside the bookshop in Mussoorie Mall proclaiming a selection of the author's books and satisfied herself with that.

Later Arun took her to eat *proper food* at the Urban Turban, he knew Mussoorie: *It is one of my most favourite towns*. The food, when it finally arrived, was good and served in an innovative collection of antique irons, spades, small buckets, but: *Oh my*, asked Anna, *what is that milky white wonder down there in the street?* Well, it was *malai doodh* and – *wondrous!* She wished she could have managed three on top of the one she had as dessert after leaving Urban Turban. The amalgam of flavours: creamy milk brought to a rolling boil in a vast concave open pan, with nutmeg and cinnamon and dates and apricots and probably ginger and definitely sugar – she could be contradictorily selective over her diet at times, Arun shook his head at *sugar* for *her* – all warmed together in that concave pan became nectar for the gods and Goddess and Sita Devi.

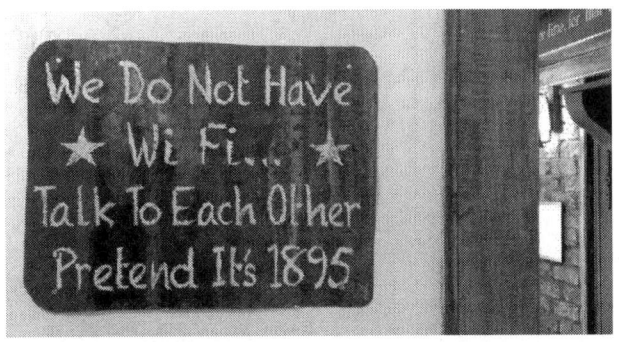

11

Miracle of the Moon

And so to Rishikesh. *Two and a half hours*, said Arun, *we will leave at nine o'clock, we will be in Rishikesh in time for your proper food.* Five hours later they reached the outskirts of Rishikesh. Anna's contingency plans for food relied either on accuracy or stocking up. With neither, the body rebelled, and the temper – well, she would cheerfully *kill*.

Traffic diversions and a missed turn left – which Anna roared at the driver to take well in advance, she read the overhead sign while he was on his mobile, *again* – *and* in the wrong lane, Arun otherwise preoccupied with his favourite body part, his own mobile phone – sent them into bedlam traffic and a two hour diversion. Anna simmered. The diversion lead them far away, and police blocks refused them entry at the next dozen left turns to Rishikesh, the only one permitted was almost at Haridwar. It took another hour to find road access to that concrete monolith of ashrams, Sivananda of Rishikesh.

Way back in Dreamtime one of Sivananda's four successors, Satyananda of Bihar, founded the ashram in Australia where Anna was initiated as a swami, a teacher of yoga with its arcane mysteries. Since her encounter with Judi-before-Dame at Charing Cross Post Office, Anna had, under the silent influence of her favourite actress, shed her dhotis. Now, she wanted return to the source of all that India had taught her. To say thank you for her journey and for all she had received; to say thank you for the riches India had gifted her, through its great tradition of yoga and metaphysics, wisdoms with which she had navigated the many losses of her own life. She shared the great Sivananda's birthday and took such synchronicities seriously, staying there would be a kind of homage.

Homage wasn't in the stars that day. She should have kept a dhoti. Instead, Arun and Anna met the rudest of old men behind the reception desk who told them there was no room at the inn. Anna hadn't even finished her sentence requesting a room. She stood, eyebrows soaring skywards, until Arun's hand on her arm pulled her away, his innate courtesy plumb-shocked at the display of ill-manners shown to an old woman: *So rude, was that man, so rude, and you an old lady! He also was old, he should show respect!*

We must eat, Arun, I need proper food, and he brought her to a restaurant he knew. Its sole customer was an asparagus thin young western woman of about the same bland colouring of the French variety of that grown-in-darkness vegetable. It didn't bode well. But Anna knew she was incapable of trawling the back lanes and streets to find something that might have more appeal. The food was good, she ate happily. Arun ate happily. He knew the place from many visits with his family. At the end of the meal the smiling waiter placed a thick notepad in front of Anna as he took away the empty plates and asked her to recommend her meal, the restaurant, the service. She was happy to oblige but first she must spend a penny, as they say.

She returned with a thunderous face. *Never,* she said to Arun, *have I seen more obscenely filthy loos,* and Arun, sitting in his customary aura of equanimity, inwardly sighed: *This is India, what is the difference?*

Anna filled in the form fulsomely deriding the filth. *Thank you, thank you,* said the smiling waiter and Anna replied acerbically, knowing the page would be binned the minute she had stepped back into the street: *You shouldn't thank me for saying this; you should clean those disgraceful toilets. You expect* me *to go in there and then to write nice things about your restaurant?*

Undaunted the waiter smiled, nodding his head as he said: *Come again tomorrow, it will be clean,* and Anna burst out laughing, all anger washed away by his insouciant response. *Darling,* she laughed, *those loos will take six months to clean. Best rebuild them! Byeeee.*

The pair left the restaurant, left the car and the driver, to walk across the bridge to a grand ashram Arun knew. But the ashram had no rooms either, bookings were made a year in advance and it was Durga Navratri. Anna began speaking of her road trip to the elegant woman at her computer, who listened, warming to Anna as an older woman, appreciating her bravery in going off alone, told her to return after four pm: *Maybe there will be a cancellation, even two. Pray,* she said, and Anna would.

Still the pair tried many other ashrams and small hotels, none had rooms; Navratri and a holy dip in Ganga Ma drew millions to Rishikesh. The pair returned to the lovely gardens of the friendly ashram, to wait in hope.

And lo and behold there were, eventually, at five o'clock, two modest rooms available.

Anna's had an open chute straight down to the kitchen below, the noise coming up from cauldrons being hurled across the fires and oven tops only outdone by the shouting from the chefs, but the room had doors, those treacherous shuttered ones of recent experience, between the bathroom and the open space with the chute and, closed, they would help baffle some of the sound. She had her good French *Quies* beeswax earplugs too, the first time on the trip she would need them. There was only one bulb working for the three compartments. Arun fixed the lights, of course, but there were no blankets and it was cold.

They took the ferry across Ma Ganga to the car to collect and carry their bags. Anna forgot her coat. Without covering she would freeze, her blood temperature dropped alarming at night. Just another bodily thing she had to factor in when she travelled. Hero made another heroic trip back across the river to collect it, bidding Anna to relax. Even so, sleep would be at a premium this night.

Supper was good. Anna paid, thalis were brought to them a few minutes later, and refilled sloppily as soon as one of the hollows emptied until a hand raised gracefully would indicate: *Enough!*

Strictly cordoned off was a section of the ghats with access to Ma Ganga owned by Parmarth Niketan under whose auspices had been built a vast statue of Siva rising from the river on its own mole, a Colossus, well muscled, calm faced, in deep meditation, pearl blue to indicate

Siva's swallowing of the world's poison. On the left of his head on this statue, it seems variable, he wore a crescent moon to signify his control over Time. Ganga, so the orthodox say, was released from the matted locks of Siva. They were *here,* Anna reminded herself, in this holiest of holy places.

Seeing the statue didn't move her a jot, though Siva as Deity did. Every evening massed aarti was performed on those private ghats, the singing of *bhajans* and waving of camphor flames, held high in oil lamps by young Brahmin priests-in-the-becoming, none of it moved Anna a jot. Even knowing she was before the Deity, *here,* with the divine element of the great river Ganga, in a spirit of humility and gratitude, receiving the blessing of the returning flame by cupping her hands and waving the flame's captured sanctity over her head, didn't move Anna a jot. Understanding the five elements being honoured: Space, Ether, *Akash*; Wind, *Vayu*; Fire, *Agni*; Water, *Jal*; Earth, *Prithvi*; failed to stir her – a jot. *Perhaps,* Anna excused herself, *I am just too exhausted.* Even the bhajans, headed by a Ma from California or Canada, failed to raise a flutter in her soul.

But, Something Wonderful was about to happen.

Supper over, Anna went back alone to the now empty ghat. Siva glowed faintly luminous in the darkening twilight. Siva had always drawn her, over the decades since she last knew him as present in her life this wondrous deity had sat in her heart quietly biding his time. Anna knew him as androgynous; in his Ardhanarishwara form the union of the god and the goddess existed as long as the living world, half female half male. It will exist that way until the end of Time proves it true. The goddess, for Anna, had more power, though she held it equally with her consort. Perhaps being a woman inclined her to think this. Riding tigers, slaying demons, hair flying, decapitating enemies, twirling her cosmic fidget spinner, smiling, alluring, a symbol of resurrection and continuity, the goddess signalled a lifeline to Anna's own beginnings and endings. It was a warm feeling sitting there with the sound of the River Goddess Ganga swirling her reveries in front of Siva. Anna's thoughts drifted to Arun, how he had done a hundred things that day to make her comfortable.

Arun needed to be with his god and goddess. He appeared silently to sit beside Anna. The pair sat close, silently, as the sky grew darker, to very dark. The crescent moon rose – and rose – and at the darkest moment of this blackest night, *travelled sideways in the sky*, to hang suspended for long, long, moments, gracing the opposite side to the crescent moon fixed to the mound of matted hair of the statue of Siva, the Colossus of Rishikesh.

The silence of that night was replete with meaning, as if Siva, shod with moonlight, was walking so close that were Arun or Anna to reach out their hands they would touch the garment of the Goddess Moon as it graced this head of Chandraśekhara. Knowing it a sure blessing on their pilgrimage Arun was moved to tears and his ever present mobile phone, body part, rose unconsciously to capture the holy moment. The image would be pixilated darkly almost out of recognition, but each of them would know forever what that blurred image recalled.

Simultaneously each whispered to the other: *If it wasn't for me you wouldn't be here watching this!* Their mutually egoic claim sent giggles along the ripples of the river and into the night wind. Later they did their own puja, letting down leaf bowls of flowers and flaming camphor and incense.

The following morning Anna looked at the statue in a new way, found it beautiful, swore Siva had made alive the holy smile on its face.

Siva, during daylight

12

Ma Ganga is Always Pure ...

We have another day, announced Arun to surprise Anna. She was not looking forward to the end of her idyllic magical-merry-go-round, nor to the drive ahead and returning to Delhi which spelled the end of the spell. Her hero was promising an unexpected day she had not accounted for. *We could stay at Haridwar for the night, have you been there?* In fact, she hadn't. Elation lifted her spirits.

The drive was uneventful. Alas the suggested timing of one hour once again extended to eternity and well beyond Anna's need for the proper food promised well in time. Was it the driver's perversity or Arun's poor memory that directed them to end up in a field of mud and dust and potholes so treacherous Anna lost her calm: *Of course this isn't the right parking place,* she snapped, *not even Indians would risk their car here,* though she could well see hundreds, maybe thousands, had. Breakfast was an age ago. She was attempting to forestall another sugar-slump – it ought to have been so easy, Haridwar was only an arm's length from Rishikesh after all the driving they had done over the past weeks. She was thoroughly, grittingly, disagreeable at that moment.

Their car was stopped as they attempted to leave, money was demanded, they had no ticketed proof of when they had arrived. Anna quietly seethed. Arun's usual honeyed charm, with a wave to indicate foreign Anna in his care, was the right password, the barrier raised and off they sped to a regulated, secured and well designated parking lot some miles further on.

The town itself was reached by a long footbridge over the Ganga and the very buzz of the air ignited Anna's heart and soul. Haridwar was splendid. The town, teemed with pilgrims for Navratri, the air was electric, filled her with well-being. Anna loved Haridwar. She felt Rishikesh to be slightly jaded and despite all the attempts of Siva and the Moon Goddess to win her over, wasn't inclined to linger there. But Haridwar! Perhaps she would love it for only twenty-four hours but it was the perfect finale to her Lavender Codicil. Her hero, for all his errant fiscal reckonings, planned to the last *pice*, she would learn later, and calculatedly over-reckoned by a few hundred thousand rupees, *lakhs* of them, had chosen brilliantly their final stopover.

Looking down as they crossed the bridge Anna's throat gagged at the filth clogging the ghats where people were bathing. Was that a dead bullock? An *arm?* The mounds of solidified pollution, rank and unmoving, almost unnerved her, despite her visceral reaction of sheer joy at being there. Even the rapid currents of the eternal river couldn't dislodge the swathe of filth blocking the lower banks. A man standing nearby felt her discomfort, sensed her unspoken western criticism, turned to her and shared a secret: *Ma Ganga is always pure, it is the rubbish people put in Her that is impure.* Anna's smile spilled out of her eyes – *of course! What a sublime answer!* She got her head around it, loved its pure enchantment and knew all was well in the World of Illusion.

There were no hotels with vacancies along the first stretch of the wide promenade, or perhaps, Anna observed as she saw expressions change when she appeared behind Arun, none that had a room for a foreigner at this special time. He motioned her to remain outside at the fourth hotel until he came out to call her in. The plan succeeded, or had Hotel Gyan friendlier staff? The cheery receptionist took them to the top of the four flights of stairs in a rickety old lift, led them round its back to another small flight and into a darkened corridor to two adjacent doors, the only doors along this hidden and sombre alcove. He opened the door to one of the rooms and Anna's heart leapt – it was *perfect!*

The hotel was built almost at the steps leading to the most important bathing ghat where thousands, hundreds of thousands, would be watching the splendid puja and aarti and prayers at sunset, and their two rooms were large, bright and airy. A wall of windows, real opening windows, looked out over the whole of Haridwar, and as long as she looked left and not right Anna could blot out the extraordinary mile length of putrefying malodorous pollution clogging poor Ma Ganga! A large double bed, thick white duvet with the prettiest pale embroidered cover and pillow slips, faced the window, and to the left of the very comfortable bed a large alcove with two sofas and a coffee table. Arun's room, with its stronger coloured linen and furnishings, a boy's room to mirror her girl's room, was equally charming.

Meet me in half an hour, said Arun, *knock on my door when you are ready; I will take you for the best lassi in India....* Anna knocked, dressed in her grey silk, overlaid with a long black sleeveless tunic and a spectacular Mondrian patterned scarf of sheerest cotton, woven in asymmetrical blocks of colour: grey, cream, black, orange, rose. Another of her Angel of the Charity Shop finds, it lit up the background monochromes of *la mode*, and wearing it made her feel good. Arun chose a beautifully pressed and folded shirt with care: pale pink with sprays of roses drawn in graphite, Anna had said she loved it when he had first worn it. *I hate blue,* he told her, *but mostly that's the colour available, sooo irr-rritating!* Anna had laughingly agreeably: *I never wear blue, so deadening to my colouring, I never cross-dress either, my fashion dictum!* They had chuckled conspiratorially, aligned in their sartorial colour-codes.

Arun welcomed her in, led her to the window and opened it wide as they stood together leaning out over the view. He held up his arm with its permanent fixture – and holding it over the five storey drop, snapped a perfect selfie of them. Haridwar had suffused their widest smiles with its magic; it was unlike anywhere Anna had ever been.

Down the stairs, into the lift, passing the smiling receptionist, into the animated air outside they went. Up the steps to the main market Arun took her, to a simple doorway flanked by a very large fridge, and a small stool beside. *Here,* he said, *sit and I will get you the best lassi, this man has been making them since I came here as a boy.* Anna could have downed two, it *was* the best, made with real rosewater. Now, time for supper before aarti; they must have eaten somewhere, Anna thought to herself as she wrote up the afternoon back in Delhi, but she couldn't think for her life where, or what. Aarti, however, was an unforgettable experience.

Arun shepherded her to the opposite side of the river, bought woven plastic seatings sold by wandering vendors specifically for pilgrims who had forgotten or neglected to bring their own, spread them in place and sat. Anna sat as close as she dared to Arun; people were arriving in their dozens, people everywhere, how many hundreds, thousands, would finally settle here to watch the spectacle of Lights and Flame on the opposite bank? The chanting began, the crush of people increased alarmingly for a Cat-Who-Walked-By-Herself, a mostly English cat at that, she didn't *do* close encounters with strangers, not even a hundred thousand *prayerful* strangers.

Yet being there was amazing. Arun could see a couple of centimetres of space around Anna and beckoned each new arrival to: *Come, come, here is room, plenty of room,* as Anna endured feet and knees in her back, a child's head on her lap, a woman's shoulder pressed unselfconsciously against her own. She breathed in its *rightness*, wishing so much that she could belong to this pulsing humanity whose focus was purely on deity and prayer. She wanted ... how much she wanted, but a memory of Tweedie, solitary, silent and grandly mysterious, reminded Anna of her own solitary Path. Humanity *en masse* with its massed prayers was not her way. She relaxed into the pulsing crowd around her, fascinated by the dusk-brought spectacle and noise on the opposite bank exactly in front of them as the maroon and gold dhoti'd priests chanted ancient scriptures, invoking the presence of the Goddess to bless with Her beneficence all who came. What a wondrous experience to be here in Haridwar, on what was nearly the last day of Navratri, with a devoted Hindu, an open sesame, Anna acknowledged, to this extraordinary holy place. Arun was in his natural element.

The night rang with bells and bhajans, the river filled with hundreds of small flickering flames, swirling away with the mighty current down towards Varanasi and the sea. An age later, aarti given and received, the great flames carried by the processional priests extinguished, Deity taken back to Her shrine, people began getting up, rubbing cramped legs and shaking down crushed and crumpled saris or dhotis, replacing sandals, picking up their seatings. Arun beckoned Anna to follow him down to the edge of the ghat, to sit on a dry step but with her feet in the water while he bought them leaf bowls of marigolds, lit the candles and the camphor with the incense, to place with prayers their own offerings to Ganga Ma. As the leaf boats sped away in the fast current, Anna reflected contentedly on a vagrant possibility: the *frisson* of guilt Arun might be nurturing over his intended fleecing her of a fee far beyond his quote!

A little girl appeared carrying a tray with brass pots of red turmeric kumkum and sandalwood chandan, her shabby blue and white satin dress liberally covered with their coloured drips and drops. She stood in front of Anna, her wooden applicator poised until Arun placed coins into her tray and she expertly daubed both pastes on Anna's upturned forehead. Anna loved the cool feeling, was conscious of a curious energy sweeping through her, very soothing, very *belonging;* she would take care not to wash it off when she went to bed. For the rarest of moments she felt one with the humanity around her, she didn't end at her own skin, boundaries had blurred; the *tilak* imprinted on her forehead the mark of divinity on the human plane. For the married woman the sign, to the enlightened mind, signifies she has been invested with the holy power of transmuting *spirit-matter*, Shakti.

The morning brought their final day together. Arun woke early, excited, today he would be with his wife and children again. It had been a wonderful interlude, Anna was special, he reflected on the places he had taken her, places she had taken him, places new to them both, touched by the spiritual integrity this woman brought to everything. He thought of the money he had earned, would make, and shook his head only a little at Anna's comment: *Money isn't everything, Arun,* when he happened to say: *Hooray, in two days I get a lot of money from you!* He'd selectively forgotten the £400 she had already given him.

Anna flinched when he had said it. She had already given him well in excess of the amount he had initially WhatsApp'd. She had paid all his hotels and food, had withdrawn rupees in ten thousands to give to him when he asked for more along the journey. She didn't keep a tally; it would have done her head in. As long as there was money in her account — and she only knew that when the ATM's spat out her input along with the debit card — she would not allow other consequences or considerations to sully her days. Was she really only a Walking Western Wallet? *No,* she said firmly to Self: *and even if I was, no-one else would have looked after me with such attention, such care, found me food along the most challenging of roads! Most of all Arun has given me a Story; a Lavender Codicil to hold to my heart for the rest of my days. He thought for me, fought for me,* remembering the Tibetans, *bought for me,* touching the earrings he had bought and had worn daily since, *protected me, fetched and carried for me, bandaged and Betadine'd my hand — oh no, I could not have had better, he was the best of men. You are old Sita Devi ji, and you've been poor your whole life, if giving him more than he quoted and more than you accounted for brings happiness to a not-so-wealthy Indian family — then give it, let it go, and hold the Story he has given you as your treasure chest.*

There would be more to treasure in the near future, a gift Arun had left with her beyond either of their knowing of it.

Arun knocked on her door: *Come,* he called, *we must go to Ganga early.* First a chai stop in the street to warm them, then up to the bridge. *Wait here,* he told her after a few minutes, and went back to the hotel. It was cold standing there and a chill river wind was curling the dhotis of the swamis standing and sitting along the broad bridge. Arun hadn't indicated how long his absence might extend, or why he went, but now Anna had cold as well as early morning pre-break-fast hunger to consider.

She started back down the steps, would at least collect her woollen shawl to cover the flimsy Mondrian she had come out wearing. She reached their floor to meet Arun on his way back down, a huge folded white towel tucked under his armpit. He waited while she gathered her shawl, together they returned to the ghats. *You can photograph me,* said Arun, *I will share them with Mamta.*

Anna did as she was bid, took dozens as he went in, splashed, ducked himself three times, came up, prayed, held on to the chains for balance against the current, chatted to two other male bathers on the same chain, separated to pray again, came out, climbed the steps, and: *How beautiful he is,* Anna gasped, as desire suddenly threatened her equanimity, *his golden colour, his so unwestern lack of self-consciousness, his lack of physical shyness, his chubby untonedness;* she ached, momentarily, felt her breath quicken, whispered: *Holy Durga!* as he wrapped himself in his voluminous towel just feet away from her, dried his body selectively to pull on clothes, selectively, dried more and dressed more, shifting the towel accordingly, frowning at Anna as she clicked away recording every movement. He had asked her to photograph him in the Ganga, praying, but *dressing*.... Seeing his frown she stopped. It seemed *dressing* was definitely not part of the photo-shoot. The rising sun lit strands of silver in his wet hair, over all their days together she had not noticed them before, her heart lurched, *lurched*, and her knees almost buckled. A memory surged up from depths she could barely fathom: *This is how I knew him.* He had been older then, silver-haired, and she had been younger, they had been together... his hair... she *knew* it; *silver*.

Arun finished dressing. He was annoyed Anna had been insensitive to his personal toilet, no one took photos then; she had watched a beautiful young woman in a deep rose-pink sari bathe and pray, her husband photographed her a hundred times in the water of the holy Ganga but the moment she set foot on the steps he had put down the camera. Her dressing would be done discretely behind her towel, respected and private. Arun was surprised Anna didn't know this. She should. He knew they had known each other before, he could remember her, young, other lifetimes, she was Hindustani. The protocol should be in her cellular memory...

Something *other* had charged this day. It would be months before Arun would say frankly over the phone when she was home, that he *loved* her, for their brief interlude had given him much. But *this* moment, when he was annoyed, had brought up something other. Anna, too dismayed at having upset him, too discomforted by the deep impact of her own desire, knew nothing of his thoughts.

They walked back, he led her along a different way, not over the bridge but on the under-path; the sun warmed them, melting the momentary discomfort. Arun took three selfies of them in that morning light: golden, shadowed, beautiful. Anna looked so young, was it possible? With his towel wrapped about his neck and the sunlight glancing through his amber eyes, his mouth soft, brooding, uncommunicative, the river wind ruffling her hair and her photosensitive glasses not yet darkened against the light, it was a photo to treasure of that moment with all its power and Far Memory.

He found a way to reach the broad promenade of the main town, led them to a seating area of four tables crushed together up a small flight of marble steps behind the best lassi in India. She was fighting a forest of feelings inside her, a mixed forest of closure and loss and endings and ... she ate what he ordered, drank another lassi, they didn't speak, he rose hurriedly, they returned to the hotel.

Anna packed. She could hear the water in Arun's room, knew he was showering, waited before knocking to call through the door that she would wait for him downstairs. He had pre-booked a rickshaw to take them back across the footpath to their driver and the car. Anna was settling their bills, another five thousand rupees, India was proving costly in this decade, when Arun arrived. The two climbed in to the waiting rickshaw, he put their two bags on the back of the wobbly contraption and the gaunt stringy-muscled rickshaw driver struggled to pedal them as far as the footbridge from whence they walked.

Back at the car Arun expressed his annoyance at the driver, a lazy young man it must be said, who had left the remains of Anna's pawpaw on the back seat and a box of now stale biscuits on the floor. His only duties were to keep the car clean and drive. Arun fed the pawpaw to a cow, put the broken puff pastry biscuits into pigeon holes along the wall by the wasteland. Everything he did, Anna loved him the more. *Harden the heart*, she had heard people say, but how was such a thing possible? She would have to learn and swiftly. She was too old to carry pieces of heart around anymore, hers had to be whole, and hers alone. She'd work on it, during the two hundred kilometres to Delhi.

13

Delhi Daze

Palace Heights was a comforting re-entry to Delhi and reality. Given probably the best room, Anna appreciated the aesthetics of a maroon and white ikat bed throw and the practicalities of the air-con unit in the ceiling by the alcove wardrobe and not over the bed where it would chill her sleep.

Came now the reckoning moment. Anna would question Arun's math as costs had risen in a vertically stellar trajectory from his original WhatsApp text; the kilometres he accounted for nearly tripled those she had seen registered on the odometer. She knew she had fallen under his spell. *It's my magic,* he had laughingly said on more than many occasions. She reflected on the gold he could charm from his grandmother's teeth. But there could be no price on such a pilgrimage and she marvelled quietly that his answers to her mild questioning went swings and roundabouts inconclusively, covering the entire sub-continent to arrive at an argument so discombobulated and a figure so disproportionate she smiled and – gave all the money he asked and an amount more besides. She reminded herself that, of course, she paid for his room in the same hotels as she stayed, and all his meals so he would eat with her, and she prayed for Dutch courage to state what she felt she should.

Arun, in his eagerness, now asked for a further twenty-five thousand rupees to the original thirty-eight thousand she had given him the night before leaving, right there in the miniscule lobby. Anna felt acutely discomforted. The reception staff, an arm's length away, watched him re-count every note she handed him. That jarred – surely by now he would trust her to give correctly and he could have slipped it discreetly into his wallet? Perhaps it was normal Indian behaviour, but she would have so preferred handing it to him in an anonymous coffee shop over ossam coffee. Dutch courage paled and failed as she meekly questioned the kilometres: *But Arun, we only travelled one thousand five hundred and seventy-two kilometres, not three thousand five hundred that you have charged me for, so why do you ask for two thousand extra at fifteen rupees each kilometre? That's thirty thousand rupees extra, over £300? And*, she reminded him, *we spent five days in Nainital not going anywhere! We could have driven the round the sub-continent of India on the amount you are now saying...* She didn't mention the initial sum she had given, nor the tantalisingly low costs he quoted.

Arun dismissed her reasoning. He had said on the first evening she arrived that they would travel three thousand five hundred kilometres and she had agreed. So it is that she must pay. Her argument did not fit his reckoning. As for his WhatsApp message, what had that to do with things? He responded with his own convoluted reasoning, increasing Anna's discomfort in the tiny foyer observed by all and sundry. She told herself: *Enough, Sita Devi ji, give it and be quiet.*

Arun left, embracing Anna with the warmest of smiles that almost, but didn't quite, melt her as he promised he would come by tomorrow at two o'clock. As he drove back to his patiently waiting wife his happiness enfolded his heart and his wallet. Mamta would welcome him. He would buy gifts, a family wedding was coming up, she would wear a Benares sari of richest crimson silk with wide borders of patterned gold thread, and bangles of gold and of glass up to her elbows. There was *Kartik ki Chauth in a few days too, the* most important and difficult fast observed by married Hindu women for the long life of their husband, Mamta's husband, Arun himself. He wanted long life that's for certain, and the Ninja ZX, he'd sat on in a showroom, took a selfie and then sent to Anna before their road trip. She'd sent him back a photo of an old American cartoon of kids looking at a bike from the 'fifties. She didn't get it. Mamta took the fast to pray for his longevity very seriously, beginning it before sunrise and ending only after offering prayers and worshiping the moon at night. Not even a drop of water would she take, and of course she would need new and special saris for that, a red or pink *lehenga-choli* with gold woven *zari* patterns. These were the most auspicious, and the most costly. This year he would celebrate in style. He thanked God. Those two thousand undriven kilometres didn't trouble him at all.

Anna went to her room, she felt bereft. *Bereft*. Her three weeks had spanned galaxies; India had that effect on Time. The intensity of her lavender pilgrimage was over. For three score years and twelve she had been pretty well alone, and here she was, dazed, in Delhi. For three weeks she had handed over to Arun every decision within her Itinerary of Dreams: he had thought for her, fought for her, bought for her, had remembered her camera when she walked off and left it on a wall, folded her shawls, cut up her papaya, taken her places she could never have found, or entered, alone – golly, what was she to do now? She had to shift through lavender to melancholy and out again to … *Anna!* And pretty damn quick. In three days time she would board a plane to re-enter the world a world away, to return to *le-chat-qui-a-marché-par-elle-même*. Again.

She showered in the marble bathroom, the powerful jets of hot water, bliss, her body drank up the ayurvedic body lotion thoughtfully supplied by the hotel and she dressed in another now well-worn combination with a fresh silk scarf and … *what now?* Her heart hadn't quite let go of being cosseted and cared for but head told her firmly: *It's Over, Get Out There.*

The ghastliest transition Anna could think of in her sensitive state was going to Chandni Chowk. The hotel lift brought her smile back, she remembered its overhead mirror, tilted her head to look at her reflection up there, it granted the Best Selfie ever – no flabby chins! Downstairs the hotel doorman bargained a price with an auto-rickshaw driver – all the staff became solicitous when they heard she had been on a Solo Road Trip. *Mad*, she knew they thought, especially having witnessed her being thoroughly fleeced in the foyer, *She Needs Looking After. Or,* could she think so churlishly? *perhaps they are standing in hopeful and metaphorical lines waiting for the benefice of my western wallet….*

It was Rush Hour Chaos. An hour and a half later she stood crushed in crowds so thick she could barely breathe.

Anna hated Chandni Chowk. She didn't often hate but this crush and cacophony was too cruel for her rarefied senses. No auto-rickshaw would take her back for less than a large amount of money, they knew it would take too *too* long in the gridlocked traffic. CP was quite a distance in even normal traffic. Anna knew they were right. Sighing, she prepared herself – in-between being sideswiped by a cow attempting to cross the bedlam and three horn-hooting motor-scooters plus a Mercedes that seemed obscenely out of place, as well as being firmly wedged between the two auto-rickshaw drivers she was failing in her attempts to bargain with – for a small adventure, the last thing she felt capable of, but needs must. *Metro, home,* she said to Self, and began to ask directions. Well, the chasm between M-e-tro said in English English and M'troo said in Indian English was unbridgeable, no one could understand her and she couldn't understand why.

Anna felt a rise of concern. She wrote the word on her hand, no paper being to hand, showed it to a man in the crowd and magically the word was recognised. He gave her directions, pointing to quite some way off. It was not easy to locate, but repeatedly showing the written word worked, street by alley, pothole by cowpat.

The station was huge, cavernous, and Anna watched a while to see How to Survive. *Buy a token,* she noted, *join the queue, ask for Connaught Place, CP, single.* The response was: *Rsheecho* and a plastic token passed across the cubicle counter. Anna blinked and took the token. She followed Ladies Security through the scans and the bag frisking (Indian ladies carry Copious Bags) and into an Arena. Here she faltered. She asked again for CP, yes, she was told, and pointed in the direction, heard that word *Rsheecho*. Puzzled and trusting she continued asking for CP until a charming older couple, speaking excellent English, it was a second language for millions of Indians, took her under their wings to a high glass wall which slid open when the now welcome train arrived. It was, she observed, a highly advanced Metro system. Her saviours came aboard with her, they were going beyond CP but would put Anna down at *Rsheecho*. She knew she was not hearing clearly, but quite soon she arrived: *Rajiv Chowk! Rsheecho!* CP no longer written under its British name!

Anna was momentarily dumbfounded by the complexity of the station and finding her way across the hugest bridge to exit at D Block was another challenge. But she did it – and there was Palace Heights. And more miraculously – *Anna!* Navigating All The Difficulties had brought her back to stand in her own shoes, on her own feet and independent of – *himself.*

By breakfast the following morning she had reassembled the erratic sugar levels of the past weeks. It was good food in Zaffran, and plenty of it, including excellent thick yoghurt and French Dalfour conserves, *sans* sugar, and as she ate the wild eagle who surprised her last March came to sit on the balcony. Wild eagles in this vast metropolis, oh incredible India. When Arun arrived she told him of her adventure of the previous afternoon: *Alone? Alone? You went all alone to Chandni Chowk? And alone by Metro...?* Disbelief underscored his response.

Yes, said Anna grinning hugely to hide her heartbeat at the sight of him; she knew he might hear it for what it was. She forbore from adding: *And back to Me!*

Her last days in Delhi awoke that subtle melancholy the heart feels when something, or someone, special has moved away from the binary circuit to leave only the vacuum of *solitary*. Distractions veiled the deeper verities but: *You will come to my home tonight,* said Arun, *to meet my family,* was a good enough distraction to hold for her last day.

The pair spent the afternoon at the offices of WWF where Anna took out membership of two years for herself and for Arun's daughter. She considered the stipulated donation of roughly £100 her way of helping Tiger or any other wild creature needing help in India. Arun looked over her shoulder as she filled in the forms: *And what is your birthday?* he said. *8.9.46* Anna relied, surprised, *why? So what is this?* he pointed to 8.9.2018. Anna knew then, laughing, that she was on the slippery slope to meltdown. She invited him to choose gifts from the shop down the stairs from the colourful forest-themed Reception and came away lighter in heart and purse. She couldn't help Ustad or Avni but hoped her little bit would add to helping other tigers.

It was almost dark when they came out, neither Anna nor Arun quite knew what to do, perhaps Anna wanted see that romantic garden again? He drove swiftly; when they arrived the gates were to close in ten minutes. A couple of quick snaps ... just as well really, meltdown would have spoilt things, thought Anna, quite dreading the finality of it all.

The invitation to share the simplicity of his living humbled her, one main room, with a terrace overlooking the tiny crowded lane added extra space, but where did their two children sleep? She sat quietly watching as Arun opened the tin door of a floor to ceiling cupboard, rather like a locker, pulled out an orange Tshirt, took off his pristine visiting shirt, folded it precisely, placing it inside. The back of his vest lacked a quarter of itself, held in shape by its ribbing. He pulled the Tshirt over what was left. Dressed in orange, with the printed face of Siva prominently across his chest, Arun sat down at a tiny, low table in the corner of the room behind the arm of the sofa and began to prepare the evening puja. Oil, lamps, incense, kumkum. All were in place when Mamta came in from the kitchen to sit as her husband read from a well-handled booklet of prayers, the timbre of his chant deeply attractive, deeply focused and deeply devoted. Once the prayers were complete he rose to let Mamta into his place on the floor to anoint the deities with the red paste: Durga; Radha and Krishna; Sita and Rama. When she rose to return to her preparations in the kitchen Anna spontaneously sat in the vacant place, grateful to add her own prayers of thanksgiving for such a journey, a few whispered Hail Mary's was all she could summon, but she knew *She* knew her intentions. Arun was moved Anna would do this; he quietly picked up her camera from the bed and photographed her in his home, at his puja, doing exactly as he had done. She really was a Hindustani, her natural action allayed his momentary doubt in inviting her to be part of his small and perfect family in the little home he loved.

Mamta laid newspaper on the coverlet, brought in a thali of food; the meal was served on the double bed in the room with no windows. She had sat on the sofa to chop the vegetables, brought in from a kitchen only marginally bigger than Anna's but seeing the couple's love for each other left her in no doubt that Arun's life was infinitely richer than her own. The personal invitation to share an evening meal with his family helped soften the parting. *You are family*, he told her. She wanted believed it.

A very happy homecoming

14

Carp Whisperer

Back at the hotel sleep was ragged that last night, a disco exploded through the walls from somewhere along the street below, Anna lay focusing on the practicalities she needed to address in the morning: complete the packing of her small case, gather a couple of snacks from breakfast for the hours in the airport, run round to Croma as soon as they opened to buy a 16GB pen drive to quickly (well, it would take an hour at least) upload all her camera photos to give to Arun. Candy needed to remain outside the cabin case to go through security, no extra packing necessary. The other photos, a total of two thousand, had been WhatsApp'd from each to the other from their phones along the way. Anna loved the fun of the phone: *How*, she smiled to Self, *could she ever have been so cell-phone phobic?* The disco stopped at three, she drifted off to sleep.

And then, it was morning. Everything achieved, time to go to the airport. Her phone rang. Arun was downstairs waiting. Last goodbyes, a final glancing grin at herself in the lift's mirrored ceiling – hmmm, last night's tilak still in place, she would land in Heathrow silently announcing where she had come from – and she stepped out of the door to find – Mohammed! Oh my, how she wished she had read his text. He had come to see her, the tuk tuk driver from last March, the street-wise young man who had chided her: *Grandmother, you must not be travelling alone without a mobile phone!* He had given her his number then, and his facebook link, and stayed loyal. Now she had no time for him at all. She gave him a hug. More than she'd felt possible with a Hindu, with Arun; Mohammed was Muslim. Words of apology tumbled from her, he had been waiting downstairs since his call – and there was Arun, by his old taxi, watching, puzzled. A tiny clearing in her confusion brought a now habitual response: *Photograph!* Arun had watched her odd performance of familiarity, what *is* she doing, she wants a photograph? Naturally he would oblige, and came over to take her camera. She and this auto-rickshaw driver stood by the battered yellow and green vehicle, snap taken. Anna would WhatsApp it to Mohammed once she was home, when she had blue-toothed it from her camera to her phone. *God!* she grinned at the back of her mind, *she was such a techno whizz, after how many weeks?*

The rest of the family were in the back of the car. Anna climbed in the front with Arun, waved to Mohammed, who watched until the car drove out of sight.

Mamta reached over and gave her a blanket wrapped in a box too large to carry. Anna asked if she could leave them the box with its proper handles, a regret, but one carryon was all she was permitted. When she pulled out the blanket she smiled with pleasure at the colour choice. Yesterday she had bought a kurta to travel home in, black with a wide gold and maroon edge on the hem and sleeves, a wide row of fawn paisley woven between the bands. The blanket was maroon, with black and fawn and ochre ikat – she turned in delight to thank

Mamta, Arun. Really, how could Arun have known she had bought the same colour coordinated garment? He read her, that man; and his wife, well she was blessed with a husband like him. Now a tiny silver pendant of Durga on her tiger was being pressed into Anna's hand too, and Arun gave her two Amul Lassi's to travel with, rose water flavoured. A last selfie of them all at the airport concourse, then Arun told the family to wait in the car, he alone would walk Anna to the doors. Then – *oh my*, floods of biblical proportions rivered down Anna's face. No sound, no sobs, just awash with tears. Arun's eyes filled, the two stood momentarily unsure, just as they had at their last goodbye six months before. He touched her hand, a half hug, Anna thanking him once more for, well, everything, as her face required his snow white hankie to wipe dry of tears. Anna turned toward the queue, Arun blew her a kiss. She was in another world as the great glass doors separated them.

Inconsolable. She was inconsolable. In the Indira Gandhi Airport she sat down and wept. She drank one of the lassi's, mopped her eyes, limped across to the endless check-in counters, asked for special assistance – fairly essential these days to avoid the stress which produced the cortisol that sent insulin soaring and caused Anna's collapses in unsuitable places. No Arun anymore to find *proper food*, but his gift of Amul with rose water was a kind of love. Anna continued to weep, copiously, silently. None of the Indians around her noticed her stricken face, she was old and probably parting from family, lots of them were weeping too.

Somewhere above Yerevan Anna discovered Gone With The Wind on the small screen. It carried her

All The Way Home. And – much laughter – only as the aircraft taxi'd to its final halt at Heathrow and the seatbelt signs were imminently going off did Scarlett raise *her* stricken face from that Grand Staircase to say to the world: *I'm Going Home To Tara!*

A most restorative note.

Anna spent the night at a private lodge in Hammersmith, close to the coach station. Her plane arrival missed by minutes the scheduled Berry Bus from Bay D that same evening.

The V&A was a good place to spend the interim day, she promised to send a photo of Tipu Sultan's automaton of a Tiger Eating A Dreaded British to Arun; and she would catch up with Jennifer, coming in from her afternoon stint at Crossbones Garden of Remembrance, a place mostpeople wish to forget, but ought to know. Then, home, her heart woven with more memories than a magic carpet.

On one of the bright sunny days following her homecoming, a homecoming blessed with a garden posy from Marion to welcome her, Anna took the leftover puffed rice she'd carried back from the Nanda Devi Temple in Nainital to feed to the sacred carp in Glastonbury Abbey. As she let it fall through her fingers the carp leapt up out of the limpid water to claim it; with each fall Anna's heart fell heavier – she fed them her heart. Became a Carp Whisperer.

Feeding over, Anna resisted going back to the dollshouse, too confining for her feelings now. She needed to walk, think, not think, feel her hips swing to her footfall, two miles would do it, to Street, then she'd catch the bus back; Street where she had her Moment of Revelation – *ossam*. Street was essentially a Quaker village built on the wealth of Clarks and Morlands, unimaginable wealth, contradicting their philosophy of *Poor is Best* often writ large on the notice board of their Meeting House. The words inevitably made her smile as she passed them. The good Quakers of Street barely did Jesus and they certainly didn't do Hindu Goddesses, but there to Anna's wonderment, at end of the High Street of Street in the Cat Protection Charity Shop, on a bottom shelf amongst the bric-a-brac, someone had donated a tiny technicolour statue of Durga on her Tiger, twirling her Cosmic fidget spinner; invincible Durga, embodiment of Divine Energy.

Walking home with her prize Anna thought of Arun, of her pure desire for him in that moment by the sacred river. *Of course!* light bulb moment to Self, *it wasn't Arun I ached for, it was his whole Universe; his birth-inherited Knowledge of his place in the Scheme of Things.*

Anna wanted *That*: to belong, to know marrow deep the words she had heard so long ago: *For Hindus, that same Force is resident in every temple, every shrine, every leaf, every dust mote. When a Hindu places her forehead on the stone step of even the simplest shrine, her action is a living symbol. Discover that in yourself, live in that secret power...* Arun lived in that secret power, all the Hindus she had ever known lived in that secret power. The secret power they knew as Shakti. The secret power Father Bede had dreamed as Para-Shakti. It was *That*, Anna had desired. *She.*

A month went by; Arun WhatsApp'd often, affectionately, *missing* her. Anna's strategy to fill her own loss was to find the two hundred perfect photos of the three thousand the pair had taken along their journey. It took time for Anna to realize the gift beyond price that her charioteer had bequeathed, something extraordinary, more than a pilgrimage, more than a lavender codicil. It was a final teaching, and it dissolved the core trauma of her lonely childhood echoing with her mother's mantra: *you are ugly, you are fat; you are stupid.* Those words had long been banished intellectually, but there were decades of her life unrecorded by a photograph and later Anna had to steel herself at those times when faced by snap happy friends who failed to read light, observe angles, apply generosity in their snap happy zeal. Even more rarely did she feel good enough inside and out to *ask* for a photo; and then only in the company of the rare friend whom she knew 'saw', and, or, when she wore something she liked enough to want to 'see'.

Arun took thousands of photos of Anna from every which way, aware and unaware. More photos than she had ever had in her life. He never asked. His confidence in his own selfies and his complete acceptance that any angle of himself was God's angle, set Anna another exponential learning vertical: any angle *was* God's angle! However Arun took her photo *was*, in effect, an angle on the way God saw her. For Arun, as for all Hindus, Goddess/God are Ever and Forever equal, differentiated only by attribute or activity, but Equal. Ergo, if S/He made her who, then, was Anna to cringe?

When she returned home and braved long looks at the thousands of photos Arun had taken Anna came to know what she looked like at long, long last. It was good enough, because it was *Goddess-given*. Wrinkles deeply crevassed in unforgiving light, overly plump, cross-eyed in bright sun, dark shadowed under the eyes at times, at other times almost no eyes at all – *but* there are a handful where shows a passing prettiness long after prettiness had left her. Old, fleetingly beautiful, Anna fell in love with her own dear face, unknown for so long – in all angles! Her mother was wrong. After a lifetime, Arun's lasting gift to Anna was *Anna*.

Durga and Her Fidget Spinner   *Carp Whisperer*

Encore ...

I am not old... she said
I am rare.
I am the standing ovation
at the end of the play.
I am the retrospective
of my life as art
I am the hours
Connected like dots
into good sense
I am the fullness
of existing.
You think I am waiting to die...
but I am waiting to be found
I am a treasure.
I am a map.
And these wrinkles are
imprints of my journey.
Ask me
Anything.

Walk in Beauty – Samantha Reynolds

Printed in Poland
by Amazon Fulfillment
Poland Sp. z o.o., Wrocław